The Alchemy of Galvanizing

The Alchemy of Galvanizing

Art, architecture and engineering

Edited by
Iqbal Johal and Nicky Smith

Contents

Introduction
Will Alsop

Materials and colour have always fascinated me; in fact I am known as an avant-garde architect who likes unusual forms.

But this is not the whole story; architecture has always been a part of me and I always wanted to be an architect, even before I really knew what architects do. When I was six years old, I designed a house for my mother to live in, but it had to be built in New Zealand!

As an architect I am full of contradictions; can vibrant colours sit alongside my interest in galvanizing? Absolutely! The slight unpredictability, changing aesthetic and the unique texture are just a few of the reasons why. This wonderful book brings together a diverse range of interventions that galvanized steel has helped create within our landscape. And it is that interaction between the object and the landscape that fascinates me most. The examples, be it the form of a building or the extraordinary work of Antony Gormley and Sophie Ryder, help to illustrate the simple yet complex relationship between zinc and steel.

I used galvanized steel for my second commission as a working architect, a visitor centre for Cardiff Bay. We designed a flattened tube that was tethered to the ground by a latticework of galvanized steel, referred affectionately, by locals, as the 'cigar on the bay'. The centre created a focal point for the Cardiff Bay Development Corporation; at the time they were redeveloping the area and creating a freshwater lake. An important aspect of the design was the diffusion of light into the structure which we achieved via seagull-like cutouts in the wooden roof and huge open oval windows that offered magnificent views of the bay. Galvanizing the steel not only added longevity but also texture, the zinc patina with its abstract shapes and refractive properties.

The project also featured a discipline that I had learnt many years earlier from my drawing tutor, Henry Bird, who greatly influenced me during my foundation course. He gave me a brick, told me to draw it and promptly left the room. I proceeded to draw it with all its shadows. On his return he went into a rage and chastised me for destroying the vision with shading, shouting: "What is wrong with a simple line?" He insisted that I redo the drawing with line only so that I could begin to see the brick and its proportions. I drew that brick for two, three hour sessions per week, line only, for three months. Eventually, he admitted that I had mastered the brick and I was allowed to progress onto the tin can. After 18 months it was the nude model. His vision was one of economy of line and discipline. It worked.

The Observer once commented that my approach to architecture could broadly be defined by my statement: "I like people. I hope it shows." I don't see the point of architecture that simply blends in. I have done lots of work with the general public and what I hear over and over again is that people are looking for something that marks their spot on the earth's surface. Something that has an identity which they don't share with others. The trouble is, it has become harder and harder to be a free-thinking architect. Very often the principal architect spends more time filling in VAT returns than getting on with what they spent seven years training to do. It is distracting. You need time and you need relative relaxation in order to work your way into a project. If you do not do that you just end up repeating what you have already done.

Even with a bog-standard building type, I like to inject some personality, as with Palestra in south London—a top-heavy office block with jazzy glazing patterns, or the Chips apartment block in Manchester, its bright-coloured facades emblazoned with giant lettering.

With the Fawood Children's Centre in Harlesden, for example, we had to stretch both imaginative design and a tiny budget. We decided to buy the biggest, cheapest and most robust structure we could find, which was effectively a standard portal frame, mass-produced for farm buildings. This allowed us to cover as much space as possible for little money. The team sought to provide a total integration of the external and internal learning environments within a simple building enclosure. A part translucent, part solid structure, all galvanized of course, helped to create a more magical space than a conventional classroom, that would help shape the children's imagination.

Currently, I think it is a very interesting time for us architects, and consequently for engineers, or vice-versa because architecture

itself is so much more diverse than it was when I was 25. There is
no predominant style or way of making architecture. There is no
predominant methodology.

It is very disappointing when you are designing something, in the
early design stages, that you do not always start thinking about
materials. But inevitably, someone in your studio, or your client asks
what are you going to make this out of? "Erm, I'll make a list." And
it's pathetic how short the list is. "We could do it in brick, we could
do it in steel, aluminium, but that's not very sustainable." But you
can not say that anymore. Then you have varieties of steel. It's
not much, is it? I always wanted to make a building out of feathers
but never managed to do that; anyway, no doubt it would have to
be held together with lashings of galvanized steel. Architecturally
speaking, galvanizing is very open and this is what makes it a
fantastic material to work with, along with other materials of course.

PS
I've won two Galvanizing Awards and it's my favourite award to have
won. You don't just get some horrible dust catcher to go somewhere,
like a lump of wood or bit of metal. You get a hand dipped galvanized
watering can (a unique object) which is really useful. I still use my
watering can because as you get older you get more interested
in gardening. It's one of those peculiar things that happens, either
you have more time for it or you're just more interested in it. Or,
it's a way of taking your mind off that wonderful subject of architecture.
But it's a wonderful thing and it's been hanging outside my studio
at the end of my garden for sevenish years and it reminds me that
actually it's a wonderful material because it changes over time and
develops a patination—in a good way, not a bad way.

PROJECT FOCUS
Showcasing the diversity of galvanizing

Catmose Campus
EllisMiller Architects

atmose Campus in Rutland, Leicestershire, is a £23 million project consisting of a new academy to replace an existing school, alongside a wide range of community facilities. Designed from the inside out, Catmose Campus is an adaptable twenty-first century learning and community environment that can evolve over time to accommodate the multiple and changing needs of its users. As one of the last Building Schools for the Future (BSF) One School Pathfinder projects to be completed, it stands as a fine example of the achievements of the programme and provides valuable lessons for future school design.

EllisMiller Architects was commissioned by Rutland County Council, Catmose College and Galliford Try to design a multi-purpose education and community complex. As the building would form an urban edge to the town of Oakham, EllisMiller was anxious to create a visually attractive building that could act as a landmark while respecting the existing landscape. A large monolithic building was therefore rejected in favour of scaled pavilions with a softer visual impact that also create vistas that make connections between the urban and rural landscapes.

The new school for 900 students opened in 2011 representing a clear departure from traditional school design—with an emphasis on flexible and adaptable daylit spaces. The design comprises four independent 'modules' each of similar plan, separated by promenades connecting to the landscape. One module forms the sports hall, while the other three are teaching modules with large social spaces in the middle. The college buildings are based on strong axial routes centred on the entrance to the community campus facilities. Six learning bases are arranged on either side of a central atrium, which contains shared resource and support areas, including a theatre, drama studios and dining areas. The first floor of the atrium, which contains an open plan library and two courtyards, is connected to the ground floor by an area of terraced seating and accessed externally by two bridges to the north of the building. The school building is designed on a rigorous grid system, with structure and services coordinated to allow movable partitions, creating a range of room sizes.

Externally, a regular series of concrete columns form a colonnade to the south-facing elevation overlooking the main square and break the size of the pavilions down to a human scale. The building is set on a plinth, rooting it into the surrounding landscape, while upper levels use lighter-weight, light-coloured materials to soften the massing of the building when viewed against the sky and topology.

Materials are durable and high-impact to reduce the wear and tear of a school environment. A carefully-selected palette of colours has been used internally to lift key spaces and create a sense of identity for each element of the campus.

The use of robust, low-maintenance materials was central to the detailing strategy of this project. EllisMiller selected galvanized steel for use externally, structurally and to provide architectural metalwork. Galvanized structural steel elements support the tensile roof covering and uses standard steel sections to provide perimeter support and gutter/rainwater collection. Elsewhere the central 'X Factor' staircase and associated balustrade is fabricated from galvanized steel. Other innovative uses of galvanized steel are its use in conjunction with the cladding, where galvanized angles form the corner details. The use of galvanizing in its various forms around the building are an essential part of the building aesthetic, as well as a key part of ensuring that the building is long-life and low-maintenance.

2012 saw a sizeable change in the achievement of students at Catmose with the headmaster linking the improvement in results to the much-improved learning environment that the students now have.

ABOVE
The design for Catmose Campus has been informed by a clear set of architectural principles: legibility, accessibility, navigability and adaptability. A large monolithic building was therefore rejected in favour of four similarly scaled pavilions with a softer visual impact that also create vistas, making connections between the urban and rural landscapes.

An innovative network of canopies provides seamless access across two of the teaching blocks. The galvanized structural steel supports the tensile roof covering forming an understated yet defined transverse route.

Exposure
Antony Gormley

Exposure, or "The Crouching Man" as it is affectionately known, is permanently positioned on a spit of reclaimed land in Lelystad, The Netherlands. At 26 metres tall, it dwarfs its UK cousin, the *Angel of the North*, by a considerable margin. The man-made site dictated the form of the sculpture, surrounded by water with a horizon that shows no end. It is a place that did not exist 50 years ago.

Exposure invites you to investigate, to walk to it. As one approaches, the nature of the object changes. You can see it as a human form in the distance, but close up you realise its true complexity; it becomes a chaotic frame through which you can look at the sky. It weighs 60 tonnes, contains 5,400 bolts and consists of 2,000 galvanized components. Antony Gormley wanted to move away from his traditional work of solid construction to an open structure similar to electricity pylons.

The project's gestation spanned more than five years. Gormley took the first step by casting himself in plaster locked in the crouching position. Cambridge University created a geometric system that was then taken up by University College London to convert this into the final stable structure. Had Fab Ltd's 'can-do attitude' was eventually instrumental in securing the contract to construct the sculpture. The team hand-cut each piece of metal using paper templates and constructed a mock-up of the giant Meccano-like piece (or at least of its vast feet) in East Lothian, Scotland.

The full complexity of the design can be seen in the formation of 55 nodal points—these roughly correspond to the 'chakras', or 'energy points, of ancient Hindu thought. "It is a re-examination of the body as an energy system, rather than as a system of bone, muscle and skin", says Gormley. He adds: "If the *Angel of the North* is an angel that's like a ship, this is a man that's like a pylon. For me it is the transparency of the form that is the breakthrough. Rather than a massive, solid object like the *Angel of the North*, it is much more tentative; it is less about confrontation and more about reverie". Something that may be overlooked within the breathtaking scale of the project is the use of galvanizing not only to protect the steel within such a harsh environment but the way its use adds to the whole aesthetic that is achieved by Antony Gormley's vision. Antony Gormley actually envisages: "Over time, should the rising of the sea level mean that there has to be a rising of the dyke, this means that there should be a progressive burying of the work".

Looking up into the structure, the full complexity of *Exposure* can be seen in the formation of 55 nodal points—the most congested of which are at the head, throat, heart, stomach and genitals. Some of these nodes are 2.5 metres in diameter, requiring up to 27 angles to create.

ABOVE
With a height of 26 metres, *Exposure* dwarfs its UK cousin, the *Angel of the North*, by a considerable margin. An adult standing next to it can barely peek over its feet. If the crouching man was ever able to stand, it would be 100 metres tall.

OPPOSITE
The project was a challenge at every stage of its development and construction. Complex modelling software was used to help with the construction of *Exposure*. Due to the exposed location of the site, galvanizing was the only suitable option for protecting the steel sections.

Cork Civic Offices
ABK Architects

The design for the new Civic Offices for Cork City was the product of a 'Design and Build' competition set by Cork City Council in February 2004. The brief required that the successful scheme maximise the economic potential of the site while providing an administrative building of the highest quality. The building, completed in 2007, provided 9,200 square metres of office space together with 140 car park spaces at basement and rooftop levels.

The site was a surface car park located to the rear of the existing 1930s City Hall on the edge of Cork city centre. The brief emphasised the need for the new building to complement the existing historic City Hall and provide modern office and public space within a new building that had a strong identity, but should not crowd out City Hall. The three facades of the existing building provided independent access to a variety of civic functions: a concert hall, a lecture hall and the council chamber. Strategically, the new building needed to create a new administrative entrance for the council to the south, revealing the backstage elevation as the new, fourth facade to visiting public and help to connect and enhance the existing space.

The new Civic Offices, conceived as a new heart to the City Hall, engage in an existing diagonal public route across the site, forming a new covered street for the city. The space is of a scale and dimension equivalent to that of the existing concert hall.

The building form is composite, consisting of masonry and glass-and-steel volumes which, in turn, is wrapped in a double facade which modifies light and air for the internal environment. The outer screen is constructed of a fine lattice of galvanized steel flats suspended from large, galvanized steel trusses cantilevered from the roof slab. Glass sheets are fritted with a solar control pattern and clamped by galvanized steel beads to form a light, elegant facade to the city. This translucent addition was conceived as a lantern—a visible expression of the ethos of openness, transparency and accountability of the City Council.

The extension to City Hall is a living, breathing structure, with particularly advanced environmental standards and designed with an unusually high level of sustainability, controlled by means of an 'intelligent' Building Energy Management System (BEMS).

Examples of this include a geo-thermal heating system that uses the water present underground to gently heat or cool all levels of the building in winter or summer respectively.

Two freely rotating 'wind cowls' on the roof of the building provide air circulation systems. The cowls align themselves to the prevailing winds, enabling fresh air to circulate through the building via a network of ducts which, in turn, supply low velocity vents in the office floors. Return air is extracted at high level in the atrium, passing through a heat transfer system before being expelled at roof level. The building features an intelligent office lighting system, which reacts to external light levels, reducing energy use wherever possible.

Faced in Italian marble, City Hall's €35 million extension has been acknowledged extensively in both Ireland and the UK for its architectural merit; amongst its awards have been "Best Public Building" from the Royal Institute of Architects in Ireland, the "European Award" from the Royal Institute of British Architects and "Best Eco-Friendly Building Award" from the UK Civic Trust.

Completed in 2007, the competition-winning design for the new Civic Offices for Cork City Council required the building to complement the existing historic City Hall. It provides modern office and public space within a new building that has a strong identity, without overshadowing City Hall.

TOP AND BOTTOM
A double facade fronts the building of the new Cork Civic Offices. The outer screen is constructed of a fine lattice of galvanized steel flats suspended from large, galvanized steel trusses cantilevered from the roof slab. Angled leafs of fritted glass on the outer facade act as a buffer from wind and city noise. Gaps between vertical leafs allow sufficient cool air to enter in summer while, in winter, the facade generates a microclimate.

The outer glazed screen of the galvanized steel facade is patterned with ten millimetre by 200 millimetre solid fritted rectangles set one millimetre apart and graded vertically. This generates sufficient shade to reduce solar gain and provides a comfortable working environment while maximising daylight.

ABOVE

The building features an intelligent office lighting system, which reacts to external light levels, reducing energy use wherever possible. A coloured LED lighting system at each level of the prominent, west-facing glass block lights up this crystalline volume in a myriad of colours on a rotating basis.

The Eden Project
Nicholas Grimshaw & Partners

The Eden Project was established as one of the Landmark Millennium Projects, opening its doors to the public for the first time on 17 March 2001, attracting over one million visitors by June of that year. Its main theme is the story of man's dependence on plants, a subject with increasing relevance as debates about environmental sustainability have gained importance. Curved like a caterpillar through a 160 year-old exhausted china clay quarry near St Austell, in Cornwall, it was a £57 million showcase for bio-diversity.

The giant biomes cover 15 hectares; the size of 35 football pitches. The Guinness Book of Records heralds the biomes as the biggest conservatories in the world. Comprising of two chains, each with four interlinked nesting domes, they encapsulate humid tropic and warm temperate regions. The four west domes contain the equatorial, tropical rain forest climate zones. Such is the height of the Rainforest Biome that it could house the Tower of London. The four east domes have warm climate zones for Mediterranean fauna.

Building these 'lean-to greenhouses' on an uneven surface that changed shape was tricky: the architect got the idea to use 'bubbles' while washing up! Many other avenues and design concepts were explored before the final, Buckminster/Pavlov-inspired geodesic system with inflated ETFE pillows, replacing glass, was selected. The ETFE offered the advantage of being smooth enough to be self-cleaning; dirt is washed away by rain. The other main concern was that of protecting the geodesic structure from corrosion. Galvanizing was selected as it met the performance criteria. The galvanized steel frame comprises almost 4,000 joints and over 11,000 rods using the Mero three-dimensional framework system.

Eden has become more than just a memorable day out in Cornwall, as was always envisioned by the driving force behind its creation and conception, Tim Smit. Eden is a charity and social enterprise providing a schools and college programme. In January 2004, a global school gardening network was launched, helping school children create gardens where they can explore food, environment and citizenship. Families are learning to plant, grow, cook, share and eat together through the Seeds, Soup and Sarnies project and a wild play scheme reconnects young people with nature. Facilitated day trips to Eden offer socially-excluded individuals the chance to get away from everyday life and the inspiration to be able to take positive steps forward. The Eden Project hosts a weekly walking club for people with Chronic Obstructive Pulmonary Disease, to help improve their health and reduce the amount of time they spend in hospital. A unique programme offering horticulture skills at Eden's plant nursery is offering a truly life-changing experience for people whose lives have been affected by disability.

In July 2008, Tim Smit took Eden's ten millionth visitor by surprise at the ticketing hall. Only 750,000 visitors a year were expected, but by Eden's tenth birthday in 2011, almost 13 million had been welcomed since fully opening.

A garden created by hundreds of homeless people and prison inmates, part of a collaboration with Eden, won silver at the Chelsea Flower Show in May 2009. A second stunning display, the largest ever in the history of the event, scooped another silver the following year.

As expected, growth of the trees called for a pruning system to be introduced. In July 2010, a large helium balloon was permanently installed, allowing gardeners to fly around the towering canopy of the Rainforest Biome. Strapped into the sling of the balloon and tethered by a colleague, gardeners can prune fast-growing trees such as balsa and kapok, which have reached over 50 metres in height. Visitors can also take the once-in-a-lifetime trip in the Rainforest Balloon.

In January 2011, Tim Smit received news that he was to be appointed an Honorary Knight in recognition of his services to public engagement with science. Together with the public vote win in the best UK leisure attraction category at the British Travel awards, Eden has marked its place as an all-round educational, socially aware, unique and fun place to visit.

The cladding system of air-filled ETFE foil cushions was chosen for its lightweight and transparency which allows maximum UV light to filter into the domes and provide good heat insulation. More than 800 hexagon elements with a width of 1.5 metres are covered in this way.

ABOVE
The new Eden Rainforest Aerial Walkway
opened to the public in July 2013. From
paths set in the sheer cliff face, the
galvanized Walkway stretches out through
the steamy canopy, immersing visitors in
the mystery and wonder of the forest.

OPPOSITE TOP
Set in an area of 15 hectares, the size of
35 football pitches, the two giant domes
are geodesic spherical networks, creating
a modern Garden of Eden. Open for 363
days of the year, over 13 million people
have visited since its opening in 2001.

OPPOSITE BOTTOM
The Warm Temperature Biome houses
the typical plants from climatic zones
encompassing areas such as the
Mediterranean, South Africa and California.
With an internal length of approximately
150 metres and a height of 35 metres, this
is the smaller of the two biomes.

The steel load-bearing structure of the
eight biomes is extremely light, considering
its impressive size. In order to achieve
maximum luminosity, the galvanized steel
structure is a two-layer, spherically curved,
three-dimensional framework comprising
almost 4,000 joints and over 11,000 rods.

Garsington Opera Pavilion

Snell Associates

Garsington Opera was founded in 1989 by Leonard Ingrams and ran for 21 years in the gardens of Ingram's home at Garsington Manor in Oxfordshire. In 2010, the Opera reached agreement with the Getty family to move to Wormsley Park, an eighteenth century country house in Buckinghamshire. It took the company a year to raise £3.5 million from corporate members and devoted audiences to finance the move.

There were some preconditions for the new venue, mainly that it should be a temporary home for each summer season with a lease of only 15 years. The requirement for the structure to be demountable led to an entirely bespoke pavilion that was constructed from modular prefabricated elements that could be lifted by crane, minimising construction time and cost. The whole steel structure was galvanized, providing a maintenance-free, durable and corrosion-resistant finish.

The new pavilion offers superb acoustics, increased comfort and a perfect setting in which to experience opera performances of the very highest quality. The 600-seat summer pavilion design takes its cue from a traditional Japanese pavilion, in its relationship to the landscape setting.

One of the most unique aspects of the design is that it was the first auditorium to have been created using a lightweight construction of steel and fabric to provide such world-class acoustic performance standards. The walls of the auditorium are all shaped and profiled, rather like windsurfer sails, to assist in sound reflections and the roof is a double-layer fabric which reduces the loudness of any rain noise by 50 per cent or 14 decibels. Both were the subject of extensive research and development, involving mock-ups and acoustic testing to optimise the design.

As the building is modular, it is an extremely flexible structure that can be adjusted as required to suit the changing opera performances. The sliding screen track allows the outer line of the building envelope to be adjusted in a matter of minutes to respond to climatic conditions.

In 2010, Garsington Opera announced
that it had reached agreement with the
Getty family to hold the opera festival at
Wormsley Park, a lush 2,500 acre estate
and eighteenth century country house in
the Chiltern Hills of Buckinghamshire.

Traditionally an auditorium is made with outer envelopes of more heavy-weight construction such as concrete or brick. This auditorium, however, is virtually transparent offering views of the surrounding countryside. A network of galvanized steel enables the structure to be erected and dismantled within three to four weeks. The tough coating withstands the annual erection and demounting process of the structure.

The 600-seat summer pavilion design takes its cue from a traditional Japanese pavilion, such as the Katsura Palace west of Kyoto, in its relationship to the landscape setting and use of sliding screens and verandas to link it to the landscape, both visually and physically.

One of the most unique aspects of the design is that it was the first demountable auditorium to achieve such world-class acoustic performance standards. Created using lightweight galvanized steel and fabric, the temporary structure is erected for the summer season.

Lewis Glucksman Gallery
O'Donnell + Tuomey

O'Donnell + Tuomey's Lewis Glucksman Gallery, situated in the picturesque lower grounds of University College Cork (UCC), is a cultural and educational institution that promotes the research, creation and exploration of the visual arts. The gallery, which opened in October 2004, is an award-winning building that includes three floors of galleries, display spaces, lecture facilities, a riverside restaurant and gallery shop. The gallery is named in honour of one of its founding donors, Dr Lewis Glucksman, a successful investment banker and renowned philanthropist who, with his wife, generously supported cultural and educational projects in Ireland.

A wooded limestone precipice dominates the campus, once the site of an ancient Augustian abbey overlooking the River Lee's meandering south channel. The University proposed a long, low building facing the river, blocking the campus from the river and building over the landscape. O'Donnell + Tuomey, however, decided upon a design concept that contradicted this completely. The aim was to exhibit and intensify the landscape. The height of the structure would be level with the trees, almost amongst them; this would help to lift its bulk, making its footprint as small as possible. Only a single tree was removed from the site.

The lower concourse is a stone-clad concrete structure, with galvanized steel windows and screens cut into the solid plinth. The timber-clad gallery spaces are supported on a concrete structure, propped by columns, tied back to the stone-clad lift tower. A steel frame structure is built off the slab to enclose the gallery spaces. The largest gallery spaces have been formed from a complex juxtaposition of geometries, which have been constructed from timber and galvanized steel panels. The galleries are clad in sustainably sourced Angelim da Campina timber, bent around the external envelope of the gallery walls and left untreated to weather naturally.

Detailed discussions between the design team and the galvanizer resulted in the selection of steel thickness and optimum size of panels to ensure desired results were achieved. The fact that the building has at least six facades is testament to the inspired design concept.

The gallery was named "Best Public Building" in Ireland in June 2005, with over 350 works in its collection. These works are sited throughout the campus to provide the UCC community and visitors with a first-hand encounter with original works of art. The UCC Art Collection concentrates on contemporary Irish art and now features many of Ireland's most distinguished practitioners.

The Glucksman actively seeks participation and partnerships with schools and community groups through its changing programme of art workshops. These sessions introduce participants to different ways of making art through creative activities in the gallery. Internships that provide hands-on experience in assisting with exhibitions, collections and education are offered, along with art activities for seniors and children.

The gallery is the personal initiative of the
President of University College Cork, Jerry
Wrixon, who sees it as a bridge between
city and University, a place where both
student and citizen can feel at ease.
The gallery is a showcase for changing
selections from UCC's art, ethnographic
and scientific collections as well as other
visiting exhibitions.

The connection between the external and internal spaces was an important element of the design, with the aim of exhibiting and intensifying the landscape. The gallery building was inspired by the Seamus Heaney poem "Lightenings viii".

OPPOSITE

The galleries are clad in sustainably-sourced timber, bent around the external envelope of the gallery walls with galvanized steel windows and screens cut into the solid plinth. The fantastic forms that give this building at least six facades are a testament to what can be achieved by an inspired design concept.

ABOVE

Self-coloured Angelim da Campina timber, concrete with sparkling granite and fossil-bedded limestone complement the galvanized steel cladding. The site is a landscaped floodplain below the low cliff on which University College Cork sits, with only a single tree being removed from the site during its construction.

Gormley Studio
David Chipperfield Architects

Former Turner prize winner Antony Gormley has created some of the most ambitious and recognisable sculptural works of the past two decades, including the *Angel of the North*, *Quantum Cloud* and *Exposure* in The Netherlands. With the increasing interest in his work, Gormley required a space that would be large enough to construct his often huge installations, yet at the same time be somewhere intimate and personal in which he could conceive his next artistic project.

Looking to satisfy both of these requirements, David Chipperfield Architects' design attempted to create light and open spaces for the diverse and specific ways in which Gormley works. The studio, which is located amid the industrial buildings, warehouses and rail yards just north of London's King's Cross station provides studio space for drawing, painting, sculpting, welding, casting and photography. The design of the studio references and abstracts the large-scale, industrial architectural vernacular of the surrounding buildings. It is distinguished by the silhouette of its pitched roofs and its bright, but even, interior light. Located to the rear of its site, the studio building itself is approached across a large yard, left open for the assembling of larger pieces. A pair of large, reassuringly weighty galvanized steel staircases, with a cantilevered landing, connect the yard to the domestic-scale studio and office areas on the upper floor of the main building. The artist's preference for metal is referenced throughout the project with bespoke galvanized ironmongery used from the large galvanized steel windows to the even larger plate steel galvanized doors that add to the drama of the open space. Operating as both a workshop and as a pared-down, white-walled studio space, the building has become the focus for all of Gormley's artistic production.

The studio can accommodate a 40 tonne low-loader truck with 929 square metres of studio space and 1,300 square metres of external space. The important aspect of the design was balancing the spatial demands of scale and proportion between the main studio with its three central bays and the courtyard. The courtyard is reinforced with 140 tonnes of steel to make it strong enough to support any type of creation.

"It's good to be able to put down five tonnes of cast iron and know the building won't collapse" grins Antony Gormley.

The studio for artist Antony Gormley attempts to create light and open spaces for the diverse and specific ways in which he works. The double-height building provides studio space for drawing, painting, sculpting, welding, casting and photography. The simplicity of the galvanized staircases makes a simple yet bold entrance to the first floor offices.

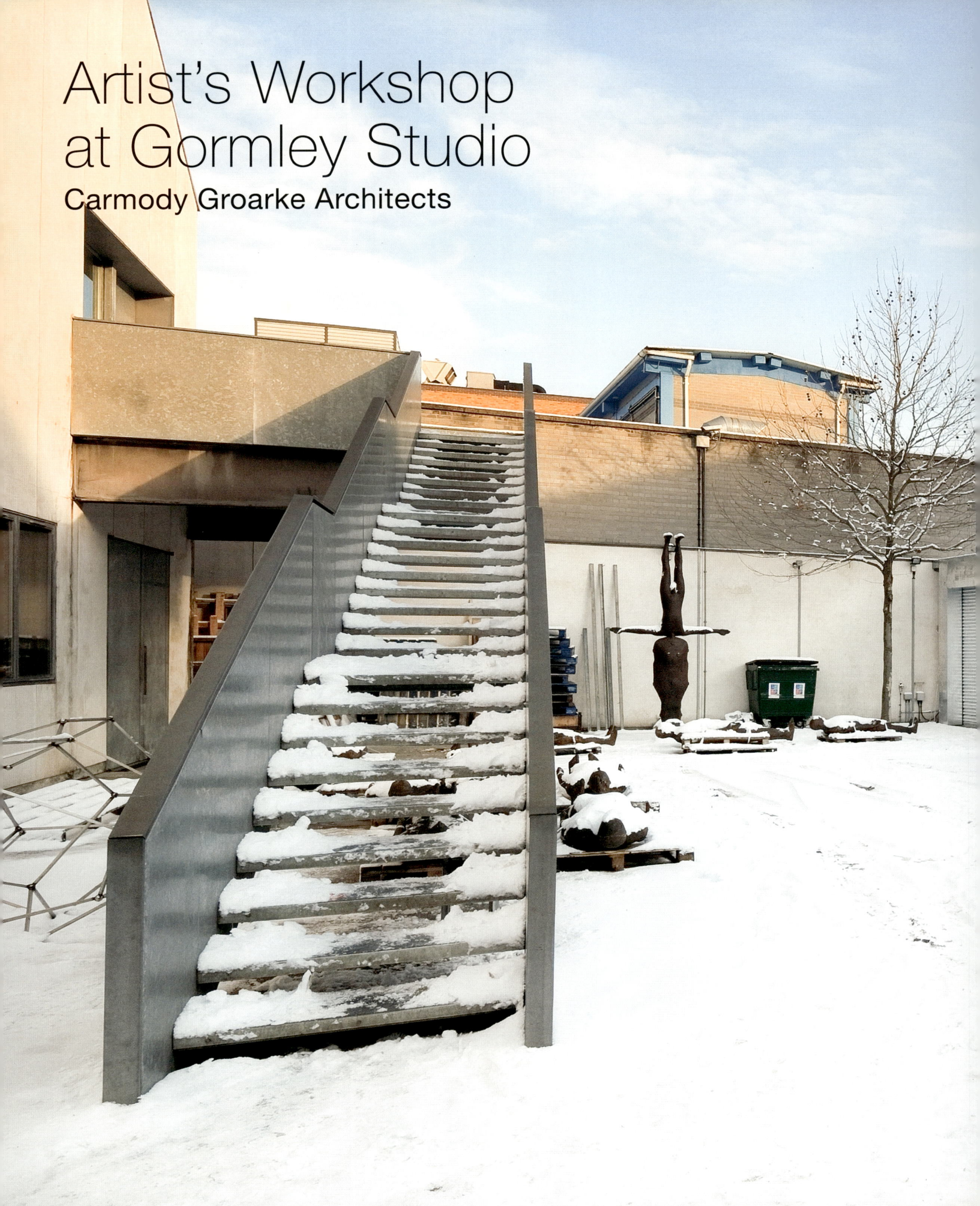

Artist's Workshop
at Gormley Studio
Carmody Groarke Architects

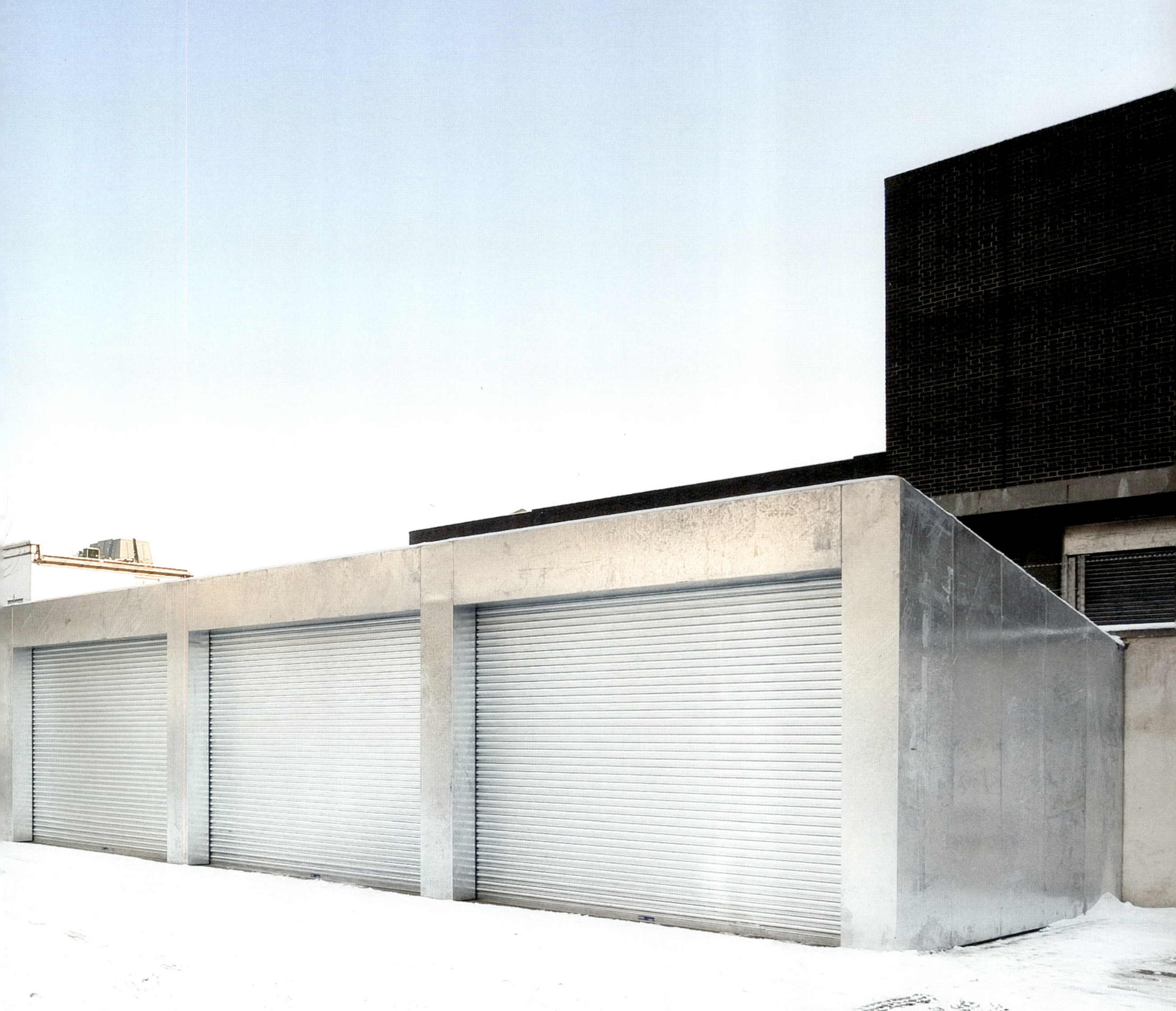

I n 2010, Carmody Groarke was approached by Antony Gormley to design a new workshop set within the yard of his London studio. As his works are predominantly made of metal, he required more storage space for raw metal materials as well as an area to carry out the heavy—sometimes noisy—work to the finishing stages of his sculptures. The workshop was consciously designed to be made predominantly out of galvanized steel, in order to withstand the industrial nature of the artist's creative process and to create a link to the existing building. The workshop has a mono-pitched roof and is split into four bays, so that the internal space can be divided as the artist's work dictates. Access into each bay is through mechanically-operated, galvanized steel roller shutter doors, which were carefully integrated into the design of the cladding and structure. The cladding was designed so as to have no visible fixings to emphasise the solidity of the new building and protect fixing points from weathering.

Considerable research and prototyping was undertaken to ensure that the process of hot dip galvanizing was controlled to achieve the desired accuracy of material junction and visual appearance. This yielded a design that maintains the protective qualities of the galvanizing to prolong the lifespan of the building and gives the building a reassuringly solid appearance.

Having created a state-of-the-art studio, Antony Gormley realised there was a need for a smaller isolated workshop where 'noisier' processes could be carried out. 2011 saw the addition of a new workshop in the studio's courtyard. The internal workspace can be split into four individual bays, each with its own mechanically-operated galvanized roller shutter door.

The new workshop reflects the natural palette of materials used for the main studio. The building's structure is made of a galvanized steel mono-pitched frame that is clad in bespoke galvanized steel panels in order to withstand the industrial nature of the artist's creative process.

Hengrove Park Leisure Centre

LA Architects

The opening of Hengrove Park Leisure Centre was a date to celebrate for thousands of local people in Bristol in February 2012. The £35 million state-of-the-art pool and leisure centre is the region's newest and most spectacular leisure facility.

The Hengrove Park Leisure Centre has been built on the site of a former airport. Its facilities include a ten lane, 50 metre international-standard swimming pool, a 20 metre teaching pool with a movable floor, sports hall, cafe and creche. Bristol City Council with contractors Kier Construction, the leisure provider Parkwood Leisure, LA Architects and Ramboll formed the team that has brought the project to fruition. The architectural design features large areas of curtain walling that is almost transparent, providing unobstructed views from the interior out on to the landscaped central plaza. The glazed facades also provide high levels of natural light and give the building a calm reflective feel at night. The entire building is formed from a galvanized steel frame with the roof supporting a network of cellular beams spanning 37.5 metres. In order to enhance light levels and create an interesting feature for swimmers, an elevated 'bubble effect' has been created over the central portion of the pool roof. The steelwork in this area is partially hidden by a suspended ceiling formed from a series of acoustic baffles that filter light and sound. In addition to the main frame, galvanized components were used extensively throughout the project. Roof plant areas are screened with louvres fixed to a galvanized structural steel framework, while galvanized purlins were used to support the metal roof decking. Lightweight metal framing was used in the construction of external and internal walls together with galvanized angle framework that supports the extensive suspended ceilings throughout the facility.

The Local Authority requirements called for the main structure to have a guaranteed life of 60 years with minimal maintenance. Traditional painting methods, utilising chlorinated rubber, were quickly discounted and the benefits of a factory-applied, robust, homogenous finish led to hot dip galvanizing being selected.

Structural engineer for the project, Ramboll, researched the benefits of using galvanizing on a wider front. The sustainability benefits gained by using galvanizing over other methods of protection were quickly realised; in fact Hengrove is the first centre in the UK with a 50 metre swimming pool to achieve a BREEAM "Excellent" rating.

HENGROVE
PARK
5

HENGROVE
PARK
3
HENGROVE
PARK

The state-of-the-art pool and leisure centre boasts a ten lane, 50 metre international-standard pool with a teaching pool and movable floor, offering swimmers the very best in public facilities. The main swimming pool is flooded with natural light from the overhead diaphanous frame created from a network of galvanized cellular beams.

Clever landscaping has transformed an old airport site into a complete leisure experience. The quirky golden dome houses the kitchen facilities for the interior cafe. The combination of the design and selection of materials has resulted in the project receiving a BREEAM "Excellent" rating.

The design concept enabled the architects to achieve a building that brings the landscaped plaza almost into the building, creating a transparent connection between the exterior and interior space. The result is a light and welcoming atmosphere throughout the building.

Imperial War Museum-North
Studio Libeskind

The first UK building to be designed by world-renowned architect Daniel Libeskind, the multi award-winning Imperial War Museum-North opened on the banks of the Manchester Ship Canal on 5 July 2002. It is the youngest of Imperial War Museum's (IWM) five branches and the first outside the south-east of England. The landmark building is a visionary symbol of the effects of war, the design based on the concept of a world shattered by conflict, a fragmented globe reassembled in three interlocking shards. These shards represent conflict on land, water and in the air.

Concrete, a network of galvanized steel tubes and a semi-porous facade combine to create the Air Shard, a dramatic symbol of entry to the museum. With its 29 metre-high viewing platform over the canal, visitors are afforded views to the skyline of Manchester. The exposed galvanized steel structure relates the museum to the former industrial harbour of the Manchester Ship Canal with the aging patina of the galvanized tubes connecting the building to its historical context. The interior of the Air Shard is exposed to external weather through open slots in the facade. Libeskind chose to use the semi-open facade in order to accentuate the interplay of changing light and weather with the network of galvanized steel to form a dramatic entrance point for the building.

In its first year of opening, 470,000 visitors passed through the doors of the charity-run museum. Permanent exhibitions are housed in the museum's first-floor main gallery space within the Earth Shard. These consist of a chronological display which runs around the gallery's 200 metre perimeter and six thematic displays in 'silos'. As part of the Earth Shard, the 3,500 square metre floor of the gallery is curved, gradually dropping away like the curvature of the earth from a nominal 'North Pole' near the gallery's entrance. A separate gallery, also within the Earth Shard, accommodates a programme of temporary exhibitions. Powerful stories reveal how war shapes lives, from a soldier's last letter home to the twisted steel of New York's World Trade Center. IWM's collections cover all aspects of twentieth and twenty-first century conflict involving Britain and the Commonwealth. Created to record the toil and sacrifice of every individual affected by war, the collections include a wide range of material, from film and oral history to works of art, large objects, and personal letters and diaries. IWM is a national repository and is accredited by the Museums, Libraries and Archives Council.

By August 2005, the museum received its millionth visitor, of which over 600,000 had engaged in a learning experience with a member of the staff. Over 100 people from across the northern regions have joined Veterans North, a membership group of proactive veterans and eyewitnesses who are willing to share their stories and experiences with the museum and visitors. Members of Veterans North determine their own level of involvement from attending special events, volunteering in the galleries, assisting in the development of learning resources, participating in learning sessions, talking to the media and enhancing the content of exhibitions.

Supporting its educational goals, the museum has operated a volunteer programme that seeks to engage local people at risk of social exclusion. Originally based on a National Vocational Qualification (NVQ), the programme was revised and relaunched in 2004. In return, volunteers work in the museum's public spaces as part of the front-of-house teams. In January 2007, the In Touch volunteer programme was launched in partnership with the Manchester Museum and supported by the Heritage Lottery Fund.

Since opening, the IWM-North has welcomed over three million visitors. Together with The Lowry and the arrival of MediaCityUK, the museum has contributed to the regeneration of The Quays area, creating a world class visitor destination for the twenty-first century.

The central structure forms the Air Shard, an important element of Studio Libeskind's design concept. The interplay of light bouncing off the network of steel within the shard structure helps to create a dramatic effect on entering the museum.

The interior of the Air Shard is exposed
to the external environment through open
slots in the facade, which allows light
to flood into the main entrance below.
The galvanized steel provides a durable
coating which also withstood the process
of site erection and handling.

The concept for the Imperial War Museum
North is a building dealing with the ravages
and implications of world conflict. It
represents a globe shattered into fragments,
reassembled as an iconic emblem of
conflict. It is a simple interlocking of three
shards, representing Earth, Air and Water
and the battles that took place there.

Leipzig Glass Hall
Ian Ritchie Architects

The spectacular Glass Hall at Leipziger Messe was part of an international competition-winning scheme that came about through the collaboration of Ian Ritchie Architects and their German partners Von Gerkan Marg und partners Architekten. Erected in 1996, the building was the largest glass hall to be built in the twentieth century: 25,000 square metres of glass supported by an imposing galvanized steel structure.

The dimensions of the central Glass Hall recall the Crystal Palace of the Great Exhibition held in Hyde Park in 1851 and it is popularly referred to as the 'Leipzig' or 'Saxon Crystal Palace'. It is an 80 metre-wide, free-spanning glass barrel structure, 244 metres long and 28 metres high. This was the first structure of this size to be built in Europe with a glass shell suspended from point fixings, with the supporting structure kept to a minimum and covering no more than 15 per cent of the surface area.

Ian Ritchie was initially approached by Von Gerkan Marg for help with the glazing of the roof and side walls. The collaboration would eventually result in Ian Ritchie Architects being commissioned to design the project to tender stage in 1993. Having closely liaised with Peter Rice, one of the pre-eminent structural engineers of the twentieth century, Ian Ritchie's team were able to combine architecture with structural engineering. The collaboration would result in some major innovations: glass to glass waterproof jointing, absence of expansion joints along the length of the grid shell, independence of the glass end-walls, inclined glass fire escape doors and robots to clean the exterior glass vault.

The proportions and combination of steel and glass succeeded in creating an external supporting framework with such a light appearance that neither the attractiveness nor the functioning of the glass shell suspended underneath was adversely affected. The structure, which has a total construction weight of approximately 2,300 tonnes, consists of a single-layer tubular structure with a grid size of approximately three metres x three metres, reinforced by a stiffening arch every 25 metres. The aesthetics of the structure required hidden bolted connections, designed as moment connections to avoid cross bracing. Because of the difficulty of ensuring an absolute hermetic seal to all the joints and bolting access holes, the inner surface of the tubular structure had to be protected, hence the importance of galvanizing the steelwork.

The breathtaking Glass Vault, inspired by
Paxton's Crystal Palace, is the heart of the
Leipziger Messe buildings. The innovative
design is 244 metres long, 80 metres wide
and 28 metres high at the apex. At such a
size it is quite remarkable that the overall
weight is just 2,300 tonnes.

A very low iron laminated glass was developed for the 6,546 glass panels which retains its transparency despite a thickness of 20 millimetres. The laminated panes of glass—approximately 1.55 metres by 3.1 metres—are each fastened under the structure at only four points and sealed with silicon profiles.

A particular feature of the Glass Hall lies in the combination of galvanized steel and glass; the two form a structural unity. The framework consists of a single-layer tubular structure (grid size 3.125 metres by 3.125 metres), reinforced by a stiffening arch every 25 metres.

The general design concept of the Leipzig Glass Hall is followed through for the access footbridges. Not only do these interlink the central hall with other areas of the exhibition complex, but they also lead to a raised viewing walkway around the internal perimeter of the hall.

OST - WEST - KONTAKTZENTRUM
EAST - WEST - CONTACT - CENTRE
CENTRE DES CONTACTS EST-OUEST
ЦЕНТР ПО КОНТАКТАМ ВОСТОК-ЗАПАД

Fawood Children's Centre

Alsop Architects

The Fawood Children's Centre in Harlesden, north London, was completed in 2004 as part of the regeneration of the Stonebridge Housing Estate. It was designed to provide a safe learning and play environment for local pre-school children. There was a restricted budget and the programme was short, which required considerable creative thinking on the part of the design team. Alsop, in conjunction with staff and key stakeholders, developed a concept which is the antithesis of the traditional approach to nursery buildings. Traditional nursery designs tend to locate accommodation within a low-level pavilion building surrounded by an area of open play space, which tends to be unusable for much of the year. At Fawood Children's Centre, the team sought to provide a total integration of the external and internal learning environments within a simple building enclosure. A part translucent and part solid roof hovers above the entire site which, together with the mesh 'walls', encloses both the open play space and the internal nursery facilities that are accommodated in separate heated enclosures. Although the outdoor spaces are not heated, they are protected from the worst of the weather and the children are free to move in and out of the nursery buildings without the need for coats and outdoor shoes.

The primary structure is a trapezoid shed enclosure, which takes the form of a galvanized steel portal frame with a deep overhanging roof, formed of a mix of opal polycarbonate roof cladding and bright pink, powder-coated, profiled steel cladding. The walls are formed of two types of steel mesh; the lower part uses a denser form to enhance the building's security, while the upper levels feature curtains of lightweight mesh that are modulated into rippling curves by elliptical coloured acrylic 'lozenges'. The internal classroom space for the nursery is provided by a number of recycled shipping containers painted in bright colours and decorated with applied artwork.

Fawood is a ray of light—especially at dusk, when the building is illuminated in a glow of rainbow colours. Inside, below the centre of the 'big top' roof, are three, three-storey clusters of brightly coloured sea containers arranged to resemble giant building blocks. Like the roof structure, they are cheap and cheerful, but they are also robust and will need little maintenance. The containers—which have their own lifts, stairs, underfloor heating and projecting balconies—are connected by walkways. "This combination of built and adapted internal environments has permitted a rapid construction programme", says Alsop. "It allows flexibility, and low-cost change in any future internal layout of the building."

The centre replaced and expanded existing community facilities on the estate by providing, under one roof, a nursery for three-to-five year-olds, nursery facilities for autistic and special needs children and a base for community education workers and consultation services.

Built within a tiny budget of £2.3 million, innovation was key. The biggest, cheapest and most robust structure was bought, which was effectively a standard portal frame, mass-produced for farm buildings. This allowed as much space as possible to be covered for little money, leaving the rest to create spaces that would have a greater impact on the children's imagination. The children have been reported to think of themselves as being at sea when they are inside the sea containers. They like to hear stories told inside the yurt, which is a far cosier and more magical space than a conventional classroom can ever be.

Butterflies
10
ten
20
twenty
30
thirty
40
forty
50
fifty
7
seven
8
eight
Maths
Area

OPPOSITE

The upper levels feature curtains of lightweight mesh that are modulated into rippling curves by elliptical coloured acrylic 'lozenges'. This not only helps to lighten one of the main elevations to the building, but it also creates a playful external facade while at the same time filtering coloured light into the building.

ABOVE

The biggest, cheapest and most robust structure possible was designed which was effectively a standard galvanized portal frame. This allowed for the most cost-effective way of covering as much space as possible for the smallest budget.

ABOVE
At Fawood Children's Centre the team sought to provide a total integration of the external and internal learning environments within a simple building enclosure which is the antithesis of the traditional approach to nursery design. A part translucent roof, together with the mesh 'walls', encloses both the open play space and the internal nursery facilities.

OPPOSITE
Three storey-high shipping containers provide an inexpensive solution for the creation of an imaginative, internal learning environment which is a far cosier and more magical space than a conventional classroom.

Roseisle Distillery
Austin-Smith:Lord

Roseisle Distillery in Elgin was Scotland's first major distillery in 30 years and, with a gross internal area of 3,000 square metres, also its largest, with a potential output of 10 million litres of malt whisky per year. The building is a modern interpretation of the traditional still house and maximises natural ventilation and daylight. The layout and massing of the building express whisky-making's three main processes: mashing, fermentation and distilling.

The mashing stage takes place in the western and central volumes of the distillery. Malted barley passes through a de-stoner and a grinding mill, then enters two mash tuns which mix it with water to produce wort. The liquid is fermented in 14, ten metre-high cylinders to produce wash, a fluid similar to weak beer. The wash then flows into the still house at the east end of the building which contains seven pairs of copper stills. The wash rises as it is heated by the stills and passes through downward-sloping tubes into condensers where it is distilled again. This completes the distillation process. For architectural reasons, the mash house is higher than necessary: a lower roof in this central area, which projects in plan, would have looked proportionally awkward. A wonderful feature of the distillery is the ability for visitors to view the eight metre-high stills through a liquid-crystal window in the conference room that can be clear or opaque.

There are a number of features that make Roseisle a distillery like no other. The sculptural design of the copper stills help to create a cathedral-like space, combining the galvanized steel frame, copper stills and the stainless steel tanks. The interplay of materials with their precise detailing help to transform what would normally be a typical process plant into a grand space that has been achieved with flair and simplicity.

The award-winning Distillery has been recognised across numerous industries and sets a new standard in building design and sustainable credentials. As one of the most environmentally-sustainable Scotch whisky distilleries, the majority of the by-products are recycled on-site in a bioenergy facility, helping the distillery to generate most of its own energy and reduce potential CO_2 emissions by approximately 13,000 tonnes through direct savings on fuel use for steam raising.

Building Information Modelling (BIM) was used to assist in the design of the facility, which was recognised by the Institution of Civil Engineers and is certified as "Excellent" under the Building Research Establishment Environmental Assessment Method (BREEAM). "The challenge at Roseisle was to design a building that wrapped around the complex process equipment required to distill malt whisky while maintaining the high level of flexibility required for future maintenance and advancement of technology", said Steve Ferguson, associate director of AECOM. "By using BIM and integrating the 3D architectural, structural and complex process equipment models, the team was able to coordinate and manage the build solution in a virtual environment prior to constructing on-site, and the result was a fully coordinated, efficient and collaborative process."

As lead consultant, AECOM created a unique, flexible structure that allows the roof and walls to be removed when equipment needs replacing. "The ceiling has structural redundancy, for when the fermentation units need to be replaced," said Nathaniel Buckingham, a senior engineer. "The client was concerned about the environmental impact of the replacement process, so we constructed a building that was able to reuse co-products from the distilling process to provide energy where, in the past, these residues would be removed and used as animal feed".

 The Alchemy of Galvanizing Art, architecture and engineering

ABOVE
The sculptural form of the building creates
a cathedral-like space, combining the
galvanized steel frame, copper stills and
the stainless steel tanks. The interplay of
materials with their precise detailing help
to transform what would normally be a
typical process plant into a grand space
that has been achieved with flair and
simplicity.

Roseisle Distillery in Elgin is Scotland's first major distillery in 30 years and, with a gross internal area of 3,000 square metres, also its largest. The building is a modern interpretation of the traditional still house and maximises natural ventilation and daylight. The layout and massing of the building express whisky-making's three main processes: mashing, fermentation and distilling.

James Leal Centre

Sarah Wigglesworth Architects

The James Leal Centre is a state-of-the-art visitor centre that replaced the vandalised Ray Park Lodge within Ray Park, Woodford Green in London. Located on elevated ground on the site of the former eighteenth century Ray House, this sustainable building acts as a gateway to the Roding Valley Corridor.

The centre stands alone among a group of trees where its angular projecting roof contrasts with the softer shapes of the park. The building's form and materials include perforated galvanized steel cladding, raw timber trusses and steel columns that provide a powerful dialogue with the natural setting. The building acts as a focal point, drawing visitors into the park where they can fully appreciate the natural landscape around them.

The visitor centre accommodates a wide range of leisure, education and community facilities in one of the area's most important open spaces. The north-facing sides of the roof have a shallow pitch and are covered in solar thermal collectors, while the south-facing sides are steeper and admit indirect natural light. An external translucent canopy defines a large forecourt to the west that acts as a gathering zone and outdoor exhibition area with cycle and disabled parking facilities. The cafe extends south over a new terrace adjacent to raised planting beds which grow edible plants for use in the cafe. In seeking to minimise its environmental impact, the centre uses earth pipes to provide cooling, roof-mounted solar thermal panels and a wood pellet boiler for heating and hot water, and natural ventilation throughout.

Ray Park boasts a vast expanse of green space. It benefits from a mature landscape and offers great space for leisurely strolls through the park and along the River Roding which runs next to the park. A Nature Conservation Team is based at the centre offering talks, events and information regarding nature and history, within the area of Redbridge. Ray Park cafe runs as a Social Enterprise, providing employment for adults with a learning disability. Training, work experience and job tasters for school and college students are available, alongside volunteer opportunities for those wishing to help out and gain experience in the field of catering and social care. The cafe also sells a range of garden and home ornaments from the Ellingham Skills and Enterprise Unit, made by adults with learning disabilities.

The James Leal Centre has proved to be a positive replacement for Ray Park Lodge. It is not only weather and vandal proof but has succeeded in attracting locals back to the area.

PREVIOUS PAGE
The galvanized steel cladding forms an integral part of the vandal-proof design; the previous centre was destroyed by fire. The brief for a sustainable building is reflected in the roof form, multiple pitches support solar thermal collectors on shallow south faces while the steeper north faces allow natural roof light.

OPPOSITE
The centre stands among trees where its angular projecting roof contrasts with the softer shapes of the park. Along with the perforated galvanized steel cladding, the raw timber trusses and steel columns provide a powerful dialogue with the natural setting.

Snowdon Visitors' Centre

Ray Hole Architects

Since 1838, the summit of Snowdon in North Wales has been graced with a building of sorts. The Clough Williams-Ellis station and cafe of 1935 reached the end of its life in 2006. There were campaigns to leave the site to its natural state but, after extensive local consultation, a competition to design the new centre was won by Ray Hole Architects, with the building finally being completed in June 2009. At 1,085 metres, this unique visitor centre is one of the highest buildings in the UK, responding to extreme climatic conditions: the mountain endures winds of 120mph, temperatures that can go below -20°C and five metres of rain every year.

The combination of extreme weather conditions and the logistical difficulties meant a collaborative effort was needed by the design team and client. Some of the design criteria that had to be considered included capital costs, lifecycle costs, projected building life and sustainable development principles. The extreme weather also meant that the summit could only be reached during the summer months. This led the design team to carry out a dry run at sea level. The structure was pre-assembled in a warehouse at Shotton where the overall design and construction techniques were perfected. This enabled the whole structure to be prefabricated and galvanized so that it could be transported to site on the existing Snowdon Mountain Railway.

Special considerations had to be taken into account for the structural frame to be designed to resist the significant wind speed and snow loadings that the building would be subjected to. In order to optimise the frame, special 3D software was used to model the building and efficiently calculate load paths. This also had to take into account the temporary stability of the frame as it was being erected in stages. Despite the difficult climate, the entire primary frame was transported to the summit and assembled in little over a week.

The entire frame was constructed from galvanized steel sections that had to be designed to resist the significant wind speed and snow loadings that the building would be subjected to. It took just over one week to transport and erect the primary frame, despite the logistical challenges and adverse weather conditions.

The final design resembles a rock-hewn bunker emerging from within the mountainside. At 1,085 metres, the visitor centre is one of the highest buildings in the United Kingdom, responding to one of the most extreme climatic locations: facing winds of 120 mph, temperatures that can go below -20°C and five metres of rain every year.

The external balustrade to the platform is constructed from galvanized steel components and designed to be easily dismantled for winter removal. If left in-situ, the weight of the horizontal sheets of wind-driven rime ice build would cause them to buckle.

Benyon Wharf
JCMT Architects

Benyon Wharf is a high-quality residential development adjacent to Kingsland Basin on the Regent's Canal in Hackney. The £9 million project provides 1,858 square metres of B1 space and 53 live/work apartments, with the retail units shielding the development from the main road. The site is a former timber yard and had been under-used for many years. The challenge was to maximise the value of the site with a high-density scheme, whilst avoiding compromise in terms of quality and privacy. The apartments are grouped around a private courtyard, allowing views through to the canal basin beyond. The majority of the apartments have a duplex layout, with an open mezzanine level overlooking a generous double-height living area.

All of the main living spaces have fully-glazed elevations, with access to private balconies or terraces. A variety of layering devices, including slatted timber screens and shutters, provide solar shading and privacy. Circulation is highly efficient with just two cores serving external access decks, which articulate the rear facades and form bridges across the openings between apartment buildings.

In the process of rejuvenating this inner city quarter, the architect and engineer (Techniker) carefully acknowledged its industrial past. The blocks of live/work accommodation are robustly detailed in brick, timber and metalwork. Glazed lift shafts, open stairwells, cantilevered walkways, bridges and balconies are all treated as secondary lightweight frames applied to the main structures. The double-height glazed screens, access walkways, staircases and bridges are all constructed of galvanized steel.

This finish lends the design a crisp edge that will last far longer than a coated alternative. The soft grey of weathered zinc takes its part in the palette of materials used on the facades with open mesh decking and balustrades on the bridges emphasising their visual lightness.

The steelwork to the walkway bridges is intricately detailed with pin-joints, rod braces and fabricated plate work. By galvanizing all the elements, problems of crevice corrosion, particularly at the joints between timber handrails and metal framing, are precluded. Over-coating would have diminished the visual appeal of fine joints and paint finishes are susceptible to tool damage during installation with subsequent patching inevitably being imperfect. The use of galvanizing allowed all the metalwork elements to be fitted without any remedial work.

The project was shortlisted for an RIBA Housing project Design Award (2004) and Design for Homes (2007), and won a Platinum Award for "innovation in urban apartment design" from *Hot Property* magazine in 2006.

Benyon Wharf is a mixed-use regeneration scheme providing 53 live/work apartments along the Regent's Canal. All of the main living spaces have access to private balconies or terraces, with slatted timber screens and shutters providing solar shading and privacy. External access is provided by a series of galvanized bridges between apartment buildings.

Acknowledging the site's industrial past, the residential blocks are robustly detailed in brick, timber and metalwork. By galvanizing all the elements, problems of crevice corrosion, particularly at the joints between timber handrails and metal framing have been avoided. The soft grey of the weathered zinc takes its part in the palette of materials used on the facades.

PROJECT ARCHIVE

Standing the test of time

Brunel Street Car Park

Birmingham City Engineers and Architects Department, 1988

Revisited by Iqbal Johal of Galvanizers Association

Sited between a very busy, part-elevated A road and a shiny new development of cafes and restaurants, near the heart of Birmingham city centre, many Brummies probably pass by Brunel Street car park without giving it a second glance. However, for some, it is a structure that deserves closer inspection.

In 1988 Brunel Street car park represented a new model for car park design; a lighter, semi-permeable structure that incorporated planters and cantilevered balconies to create a softer profile than that of its monolithic contemporaries. A car park should be an attractive building in its own right, was the philosophy of its designers, with subtle landscaping playing a crucial role. The multi-storey steel-framed design with long spans provides ramped, column-free parking for 390 cars.

2013 marks its silver anniversary: it has witnessed the rebirth of Birmingham with the construction of Symphony Hall, redevelopment of the Bull Ring, the resurrection of its central canal network and, more recently, the new Birmingham Library. In that time, it has taken all that the public can throw at it.

Construction started in November 1987 and was completed 35 weeks later in July 1988. All 350 tonnes of structural steelwork were hot dip galvanized whilst the perimeter cladding of galvanized steel mesh with a 'V' profile was powder coated in poppy red. At the time of its construction, a 20 year maintenance-free life was considered to be the target for the specifiers.

25 years on, the red wire mesh cladding still creates a colourful and interesting feature when approaching the car park. On a sunny day, reflected light helps to give the mesh an additional dimension creating both an opaque and translucent facade depending upon the angle of view.

The planter insets and balconies continue to be used and provide an interesting mix of colour and structure. Internally, the galvanized steel has developed a dark grey patina. Some structural beams look virtually as they did when they were first erected 25 years ago. In other areas the natural abstract shapes of zinc crystals within the galvanized coating have been revealed.

Over the last quarter of a century, car park design has entered a new era and the technology of parking has changed beyond recognition from the late-1980s and in this respect, many changes have taken place within Brunel Street. However, the simplicity and honesty of its original design still competes with its contemporaries.

A 25 year maintenance-free life has already been achieved at Brunel Street car park. It may not be over-optimistic to hope for at least double that.

Roof levels closed due to inclement weather
Town Hall Car Park
STAIRCASE "B"
Diversion END

    The Alchemy of Galvanizing Art, architecture and engineering

OPPOSITE BOTTOM
Some structural beams look virtually
as they did when first erected 25 years
ago. In other areas a stronger zinc patina
has formed that reveals the natural
abstract shapes of zinc crystals within the
galvanized coating.

OPPOSITE TOP AND ABOVE
25 years on, the red wire mesh cladding
still creates a colourful and interesting
feature when approaching the car park.
On a sunny day, reflected light helps to
give the mesh an additional dimension
creating both an opaque and translucent
facade depending upon the angle of view.

Homes for Change
Mills Beaumont Leavey Channon, 1996

Revisited by Charlie Baker of Urbed

omes for Change and the live/work co-op beneath it have been in this building, in Hulme, Manchester, for 17 years. It was a painful birth and the early years were difficult. Such a radical use for such a radical piece of architecture was always going to be so.

Yet it was a winner of the Housing Design Award in 1997, Best Practice in Urban Regeneration Award in 1998 and let us not forget the much sought-after Hot Dip Galvanizing Award. "It took a bunch of anarchic young Mancunians to achieve what had evaded all the nation's developers combined", Rowan Moore said of us in *Blueprint* in 1996.

We had built what was essentially a squared-off crescent of concrete decks and cross-walls while its ancestors were being demolished around it. 50 deck-access maisonettes sat on top of 1,400 square metres of workspace in an uncompromising urban block rising six storeys from the back of the pavement. Written into the *Hulme Guide to Development* as an aspiration, we had shown it could be done by uniting a forward-looking community behind what seemed such a radical goal of living in a proper piece of city—which was ours.

We were such a new Registered Social Landlord that we were not permitted to develop ourselves so we partnered with the Guinness Trust. Working well initially, it got more difficult as the Trust's lack of experience of this kind of building became apparent. That problem rippled out through their quantity surveyors and to the builders. The architects were a joint appointment and Mills Beaumont Leavey Channon did a valiant job at learning fast, having worked with us to create a beautiful proposal that was the centrepiece of the Council's stand at the Institute of Housing's conference in 1994.

The learning curve after completion of the building was just as steep as its incubation but revisiting it now, it is reassuring that affections for and loyalty to the place are still very strong, borne out by the glacial rate of turnover of properties. The management still sounds from afar as though it has not become any easier and I suspect that the 'zero tolerance' of mess obsessing so many social landlords would be challenged by the quite obvious lived-in look of the place now.

Homes for Change was supposed to be both a scaleable and replicable model—a community-based developer sensitively filling in local gaps in provision and developing components of the city. Had we not been blocked by those for whom an 'imaginectomy' is a necessary procedure, the land next door would be a thriving community of workspaces employing local people; different tenures of housing to create a proper mixed community.

In terms of the building itself, much of what was revolutionary then has been, if not normalised, at least enabled. The project prioritised extreme energy efficiency, put housing and workspace together when we were told it did not work. We put gardens replete with grass on the second and fourth floors and used cedar alongside galvanized steel—and maybe started a new trend. Car parking was reduced to about 50 per cent of what was usually required and we reintroduced deck-access housing.

The interest in the place over the years has demonstrated that there is much demand for this kind of contribution to our cities yet the developments around it seem to have no understanding of how public realm is the joint contribution of many buildings not the 'look at me' vanity of huge great scaleless office and housing blocks.

The building has mostly weathered well; the cedar has changed to a much softer aesthetic and the plentiful use of galvanized steel has lived up to its expectations of just being there and doing its job.

The use of materials, the scheme's sustainable credentials and its community ethic still offers a model for the regeneration of our cities and their social and public realms; perhaps such radical proposals take a while to prove themselves to the more cautious.

Diversion

The Hulme Estate buildings contain 50
flats and maisonettes together with 1,500
square metres of non-residential space
that is used as studios, offices, retail units,
workshops, a theatre and a cafe. These
unique buildings were designed by their
future tenants in conjunction with the
architects and developers.

One of the main design criteria was that the buildings should require very little maintenance. As a consequence, all of the external structural steelwork was hot dip galvanized. 17 years on, the galvanizing is performing to expectations.

Centro Vicente Canada Blanch School

Dols Wong Architects, 1998

Revisited by Juan Dols

O ver 15 years ago we created a unique covered area within an existing playground for a Spanish school in west London. The design aimed to insert a large-span structure within the existing trees which would be sympathetic to its surroundings and soften the rear of a dour, institutional Victorian block. The Local Authorities had designated the vicinity a conservation area and so required a structure which would relate to the scale and quality of the locality.

Funded by the Vicente Canada Blanch Charitable Foundation, set up to promote educational links between Spain and Britain and supported by the Spanish government, the school offers full bilingual education for children from the ages of six to 18. The brief for the project was to create a covered area in the existing playground which could be used as a five-a-side football pitch and as an assembly point during rain-interrupted breaks.

The final design resembled a long, covered colonade. Galvanized steel columns were used to support the trusses, composed of American Southern Pine members with purpose-cast aluminium nodes. These in turn were connected to the columns by cast steel heads from which sprung the mast and tail, used for post-tensioning. The trusses were covered with clear polycarbonate and a tensioned translucent fabric roof stretched between them completed the canopy.

Since then, the covered structure has withstood the vicissitudes of time, of many thousands of children and of many footballs too. It has become a well-loved, integral part of school life and the lynchpin of the PE department.

It has weathered well and has proved a tough, durable structure but, like many school buildings, has suffered from a lack of even the most basic care and attention. The structure was originally designed to be relatively maintenance-free. The columns, the only parts accessible to the students, were galvanized to avoid the need for maintenance. The rest of the structure, at high level, would only require tensioning every other year. However, the funding system of the school was such that the covered structure was practically left untouched for a decade and the corrosive fumes from the nearby Westway overpass were beginning to take

their toll. At this point the Spanish Ministry of Education agreed that the tensile roof was nearing the end of its life. During the school summer holiday of 2008, the fabric roof was stripped off, replaced and the wooden members of the trusses checked.

In stark contrast, all the galvanized steel sections have continued to perform well and have required no maintenance at all. The structure has served and continues to serve its purpose and is an aesthetically pleasing addition to an otherwise dour institutional building.

OPPOSITE
The existing Victorian school block had
no large indoor spaces for sporting
activities. This led to the innovative design
solution of a covered area in the existing
playground which could be used as a five-
a-side football pitch and as an assembly
point during rain-interrupted breaks.

ABOVE
The structure has become a focal point
for the school and has withstood the
vicissitudes of time, of many thousands of
children and of many footballs too. Over
the last 15 years, only the fabric roof has
needed any maintenance.

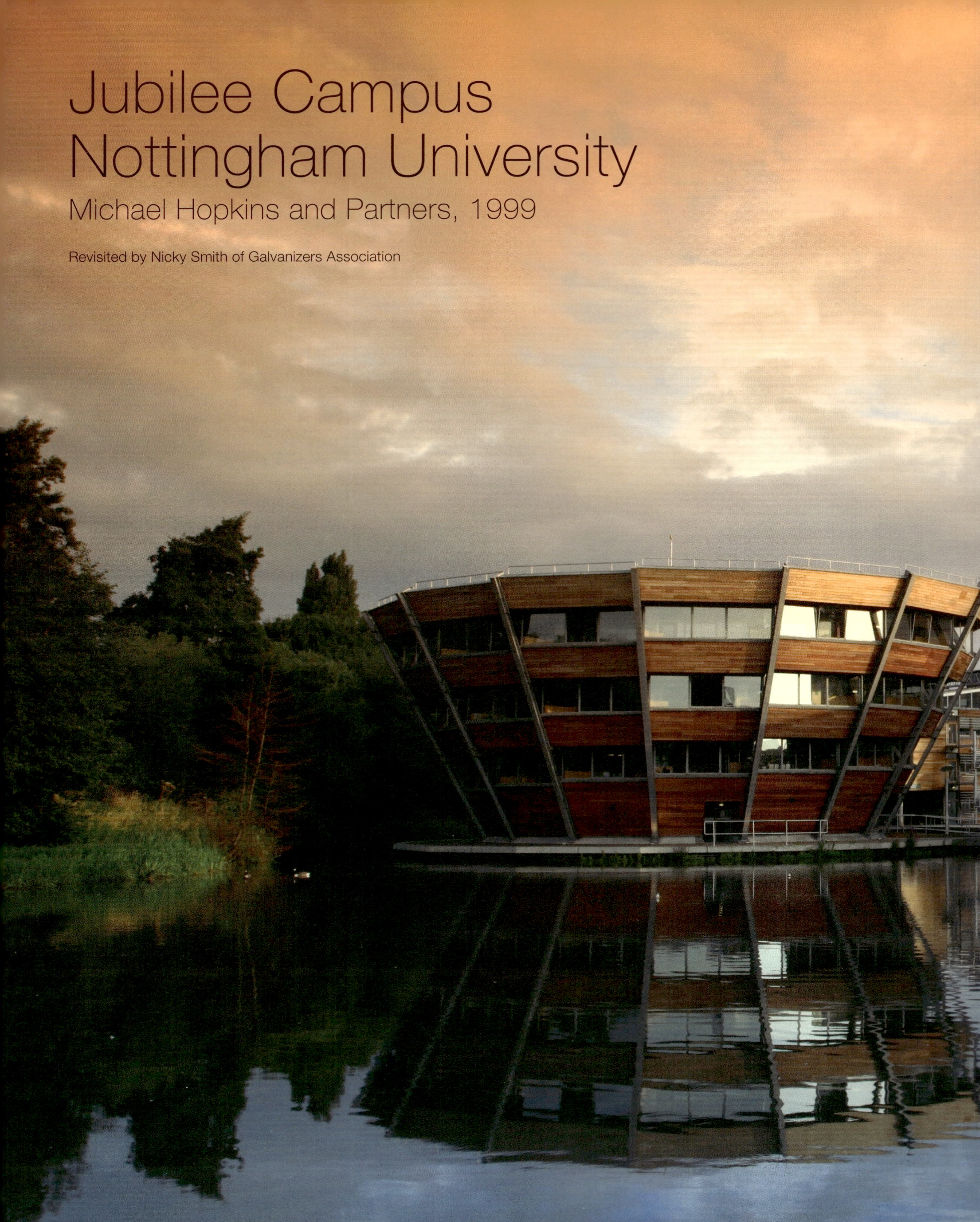

Jubilee Campus
Nottingham University
Michael Hopkins and Partners, 1999
Revisited by Nicky Smith of Galvanizers Association

The past 14 years of hard use by the students and general public have left their mark on Nottingham University's Jubilee Campus. However, the inspirational design model pursued by Michael Hopkins and Partners has risen to the challenge. During the project's construction, Jan Mackie of Michael Hopkins commented: "We chose galvanized steel over stainless steel because we wanted a material with low embodied energy, which was environmentally friendly." Jan's original aspirations have come to fruition in that, a decade and a half later, both the design concept and material choice are fairing well. The cedar has matured and portrays a rich texture alongside the patinated galvanized steel that frames it. Although not immediately obvious, the presence of gently weathered galvanized steel is also evident in many other areas throughout the campus.

The unusual shapes of the circular learning resource centre and tiered lecture halls, which appear to float in the atrium of the central teaching building, proclaim their importance. Halls of residence have more privacy; those for undergraduates adopt a traditional courtyard layout, while the postgraduate residence is crescent shaped.

The £50 million Jubilee Campus development opened in 1999, subsequently phase two opened in March 2009. The site now extends to 65 acres and state-of-the-art facilities include the University of Nottingham Innovation Park. Walking around the campus, it is easy to understand why the University campus has won numerous awards for its environmentally-friendly features.

When constructed, Jubilee Campus coincided with Nottingham University's Centennial Jubilee. The unique, free-standing circular learning resource centre and conical lecture halls made full use of their lakeside setting. The design included a sophisticated natural ventilation system for its time.

Alongside its very strong sustainability credentials, Jubilee Campus created a variety of new spaces for the students. The large glass and galvanized steel atrium continues to provide a light, open space that is used as a dining and recreational meeting point.

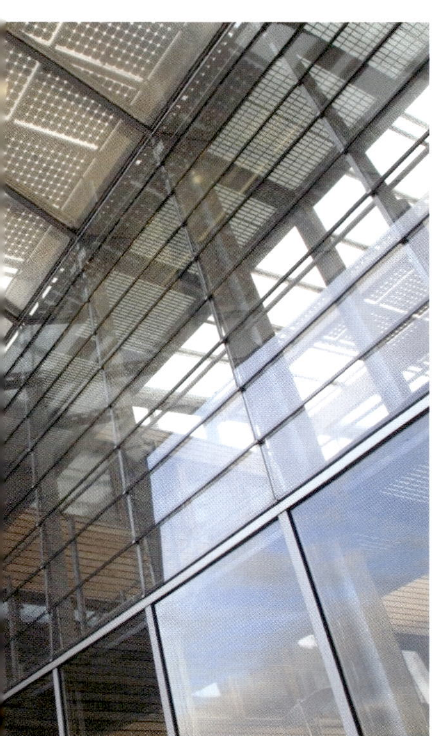

Material choice was an important aspect of the original design and timber was used alongside galvanized steel. Galvanizing continues to provide the requirements originally envisaged by the architects, who commented: "We chose galvanized steel over stainless steel because we wanted a material with low embodied energy, which was environmentally friendly."

PERSONAL PERSPECTIVES

Views expressed, visions shared

Why galvanize?

Iqbal Johal, Galvanizers Association

Ever since humans started to explore their environment there has been a long trail of discovery and invention. This includes making the most of whatever we can find in the earth's crust. Over the millennia we have become intrigued by, and ever more reliant on, metals and minerals.

Over 300 years ago, alchemists dreamt up a reason to immerse clean iron into molten zinc and, to their amazement, a shimmering sliver coating developed onto the iron. This was to become the first step in the genesis of the galvanizing process.

The story of zinc is closely interlinked with that of galvanizing. Ornaments made from alloys that contain 80 per cent zinc have been found dating as far back as 2,500 years. Brass, an alloy of copper and zinc, has been traced to at least the tenth century BC, with Judean brass found in this period containing 23 per cent zinc. However, zinc metal was not produced on a large-scale until the twelfth century in India, while the metal was unknown to Europe until the end of the sixteenth century. The earliest evidence of the use of pure zinc comes from Zawer, Rajasthan in the ninth century AD. Zinc was probably named by the alchemist Paracelsus after the German word Zinke and German chemist Andreas Marggraf is given credit for discovering pure metallic zinc in 1746.

The famous Indian medical text, Charaka Samhita, written around 500 BC, mentions a metal which, when oxidised, produced pushpanjan, also known as "philosopher's wool", thought to be zinc oxide. The text details its use as an ointment for the eyes and a treatment for open wounds. Zinc oxide is used to this day for skin conditions, in calamine creams and antiseptic ointments. From India, zinc manufacture moved to China in the seventeenth century and 1743 saw the first European zinc smelter being established in Bristol.

In 1742, a chemist named Melouin presented a paper to the French Royal Academy in which he described how a zinc coating on iron could be obtained by dipping it in molten zinc. Interest in Melouin's discovery spread quickly through scientific circles and the first application was to use molten zinc as a cheap protective coating for household utensils. These products were fairly well known in parts of France during the second half of the eighteenth century.

Today galvanized steel is all around us. Our power is supplied by galvanized pylons, our motorways are lit by galvanized lighting columns and you may have recently sat on a galvanized park bench.

In 1780, an Italian, Luigi Galvani, discovered the electrical phenomenon of the twitching of a frog's leg muscles when contacted by two dissimilar metals, namely copper and iron. Galvani incorrectly concluded that the source of the electricity was in the frog's leg. The term "galvanization" began to appear in the lexicon, connected partly to the work carried out by Michael Faraday. The metaphorical "galvanize into action" referring to suddenly stimulating a complacent person or group to take action is thought to be linked to his work.

Experiments with dissimilar metals were further pursued by Alessandro Volta, who came to believe that the flow of electrical current was caused by the contact of the dissimilar metals themselves. In 1800, Volta was able to prove this by constructing a stack of alternating zinc and silver plates with a piece of cloth soaked in a salt solution between the individual plates. This device, known as Voltaic pile, was the world's first battery.

In 1824, Sir Humphrey Davy showed that when two dissimilar metals were connected electrically and immersed in water, the corrosion of one was accelerated while the other received a degree of protection. From this work he suggested that the copper bottoms of wooden naval ships (the earliest example of practical cathodic protection) could be protected by attaching iron or zinc plates to them. When wooden hulls were superseded by iron and steel, zinc anodes were still used. In 1836, Sorel in France took out the first of numerous patents for a process of coating steel by dipping it in molten zinc after first cleaning it. He provided the process with its name "galvanizing". It is interesting to note that Sorel was aware of the electrochemical nature of corrosion and the sacrificial role of the zinc coating on the iron. In addition to Sorel's patent of 1836, a British patent for a similar process was granted in 1837 to William Crawford.

By 1850, the British galvanizing industry was using 10,000 tonnes of zinc a year for the protection of iron. This period also saw the invention of an engineered material that would help to embed 'galvanizing' into the language of people across the entire globe. In 1829 Henry Palmer of the London Dock Company was granted a patent for "indented or corrugated metallic sheets"; his discovery would have a dramatic impact on industrial design and galvanizing.

The London Dock which had only been built in 1805 was bursting with the strain of wine, spices, coffee, cocoa and wool arriving into the world's consumer capital. Henry Palmer, past assistant to Thomas Telford, was given the job of overseeing the construction of a new dock. To solve the problem of roofing massive new warehouses, he came up with the invention of lightweight corrugated iron sheets. The corrugations made the sheet more rigid so that less framing was required to support it as a roofing material. The first building to use corrugated iron was the Turpentine Shed in 1830. It was praised for its "elegance, simplicity and economy".

It was soon realised that the iron corroded quickly, this was solved over a period of time by the introduction of hot dip

Lieutenant Norman Nissen's vision of providing cheap and simple accommodation for soldiers during the First World War resulted in his concept of semi-circular steel frames supporting corrugated iron sheets. His design carried through to the Second World War with this particular Nissen hut being erected at an anti-aircraft gun site on 20 November 1944.

galvanizing the corrugated sheets. Although uncertain, the first use of galvanized corrugated iron is believed to be for the Navy at Pembroke Docks, Wales in 1844. The use of the material soon spread to the railway industry. The 212 foot roof-span of Birmingham's New Grand Central Railway Station in 1854 being the "largest hitherto attempted". Innovative bow string trusses were used in its construction and 5,945 square metres of galvanized corrugated iron sheets covered half of the roof area, the remainder being fluted glass. The other landmark station of the time, Paddington, designed by Isambard Kingdom Brunel, opened in 1854.

Brunel was still learning about galvanized corrugated iron when he started to work on the construction of Paddington. He designed a building with 36,650 square metres of galvanized corrugated iron sheets that enclosed two thirds of the vaulted roof. The sheets were used in such a way, with the corrugations at right angles to the roof, that the whole structure would stiffen. This would make the roof a very early example of a 'stressed skin' design. The great advantage of corrugated iron was now coming to the fore; its lightness offered builders the ability to build very large-spanned buildings with correspondingly light supporting structures.

The use of galvanized corrugated iron continued to gain ground with the next innovation of its use within portable buildings. These were shipped across the globe featuring in gold rushes of the 1840s in California and Australia. In fact, to this day, it still retains a certain affection in the hearts of many Australians. The boom saw the creation of the first of many galvanizing companies, some of which are still in operation today.

One further notable use of galvanized corrugated iron is during the First and Second World Wars in the form of the Nissen Hut. The idea came to an American-born engineer serving with the Royal Engineers stationed at Ypres in 1916. Having experienced problems with lack of billets for soldiers, Lieutenant Norman Nissen envisioned a semi-circular form composed of corrugated iron sheets supported on a steel frame. His design would be used throughout both World Wars, creating an indelible link in the English psyche between the material and the hardships connected with war.

The intervening time has seen iron replaced with steel and the use of this light, reusable, incombustible, recyclable, earthquake-proof material being used across the world. Corrugated sheet may even be responsible for keeping the elements from more peoples' heads than any other material ever used: a consequence of its prominence in the developing world.

Galvanized steel continues to play an important role away from the limelight given to many other processes. The use of galvanized corrugated iron spread into the agricultural industry, to be replaced by steel that is used for gates and farm buildings today.

Galvanized steel has crossed many boundaries and is used all around us today. Our power is supplied by galvanized pylons, our motorways are lit by galvanized lighting columns and you may have recently sat on a galvanized park bench.

The Lynher Dairy in Cornwall is a single-storey building with the entire frame constructed of galvanized steel. The project's inception coincided with the foot and mouth outbreak at a time when the farming industry was in chaos. The planners considered the application on the basis that the building would be as sensitive to its surroundings as possible.

Today's architects use it because of its unique properties as detailed by John Parker of ABK Architects: "Galvanizing gives texture to steel; it has a certain materiality and reflects light in a certain way not found in stainless or aluminum."

Perhaps it is the honesty of the coating—a reflection of the chemistry of the steel that it is reacting with that gives it an almost natural feel. The fact that its final aesthetic is not in the control of the specifier or galvanizer but a mixture of chemistry, steel thickness and design sets it apart from the over-engineered and inert coatings of more recent times. The fact that the coating aesthetics change over time also adds another dimension to its attributes.

Some may feel it is heresy, but today it is possible to successfully apply an organic coating, such as paint or powder, on top of galvanizing, the development of which includes the work of a dedicated British Rail engineer more than 40 years ago.

Galvanizing is a proven, honest and sustainable coating that has not lost its links to the alchemists that created it many hundreds of years ago. Even they would be surprised that it continues to go from strength to strength into the twenty-first century.

TOP

Located in a park in the mountains of Hyogo Prefecture, Japan, this public convenience gives the traditional design concept a literal twist. The architectural concept of this facility aims to form a linkage between openness and closedness through continuity of galvanized corrugated sheets.

BOTTOM

Passage is provided in three directions with no clearly defined entrance, thus avoiding defensiveness. Interior walls double as exterior ceilings and floors, the sweeping, undulating steel forming a linkage of changes, challenging the architectural norm.

An engineer reflects
Matthew Wells, Techniker

Maybe zinc deficiency causes memory loss. My earliest recollection was in 1960, aged four, of a stiff-rimmed tin bucket, left in the garden for years. The coating of zinc had failed in one place, the rusty scar, feathery and pin-holed had been plugged with a metal disk sealing on cork; post-War austerity. Underneath the bucket was an emerald green frog that suddenly disappeared with an eye-blink spring. I used to go back and look under the bucket but it never came again.

Driving along, we would see Dutch barns in the fields, turned chocolate brown and about to dissolve. Newer barns looked dullgrey and healthy. Much later I read that the galvanizing gave up after exactly 11 years (an obvious misconception). Pylons in the fields were brighter metal and, shiniest of all, were Armco motorway barriers covered in big crystals of soft metal.

When I first came to London we would sit out on the roof. For the budget flat conversions of Finsbury Park dormer sheathings, flashings and gutters were all in thin zinc (element 30). Paler grey than lead with a slightly powdery surface, sharper edged and occasionally overworked and damaged by the plumber's dolly, the metal was less permanent-looking than element 82 but homely just the same, not quite fitting around or into the corners.

I was taught the chemistry of corrosion in engineering school. As architectural students, we were shown assemblages of galvanized components and how to form sheet components. We did not encounter the metal itself. My first job found me working in a team on the composite framing of Peter Foggo's Gateway 2 Building in Basingstoke, an atrium headquarters building. Cruciform columns supporting the feature lift were assembled from rolled steel angles welded together. Straight and true into the galvanizing bath they went, out they came in fantastical helical shapes as the heat released weld distortions. A few design and fabrication adjustments sorted things out. Graduate architects get to do a lot of interior work and stylish precedents were all around. Parisian bistros—'le zinc' were sources of inspiration, bar tops and tables given a working veneer, wiped clear of beer and sauerkraut ready for the next customers.

The architecture of the time was dominated by a renewed interest in construction. High-tech was a very mannered style—always

In the United Kingdom a structure called a Dutch barn is a relatively recent agricultural development meant specifically for hay and straw storage; most examples were built from the nineteenth century. British Dutch barns represent a type of pole barn in common use today. Design styles range from fixed roof to adjustable roof; some Dutch barns have honeycombed brick walls, which provide ventilation and are decorative as well.

a demonstration of the solution in the solution itself. Better understanding between architectural and engineering professions was expected to make better solutions for the building environment. I had trained as an architect and engineer and believed that new materials truthfully deployed would make a new and timeless architecture. The profile of engineers was rising. I switched to the office of Tony Hunt. He was inquisitive and rigorous about the right way to construct things. He taught me that engineering cannot just deploy itself as a complete system, that architectural enquiry and aspiration must inform its proper use. We experimented with galvanized pressed metal, lightweight and looking 'manufactured'. The mathematics of monocoques and thin-wall structures were enjoyable applications of our aircraft structure courses. We took Ward beams, galvanized purlin rails, mass-produced, engineered down-to-the-bone, flimsy and close-fitting, using them wherever we could. Michael Hopkins' Schlumberger Research Centre, then Patera office system, mixed sheet metal panels and rail elements with rolled steel framing. As project engineer for Allies and Morrison's Herman Miller building on Pentonville Road I encountered zinc in another role. The cemetery was declared cleared but, digging foundations, out came zinc-lined coffins. The remains of small-pox victims might still have held viable strains of the disease now eradicated by the World Health Organisation.

Designers in my early years sought an integration of different inputs into building design. The architect would lead structural engineer, mechanical and electrical consultant and surveyor in an orchestrated process that would produce well-balanced buildings. For the technologist a sensitivity to architectural enquiry as well as an understanding of the 'right' use of a material or system was the ideal. Engineers would look after their speciality and let it exert its influence on the developing design, always acknowledging the architect's central role. As time has passed, emphases have changed. Sub-contractor influences have promoted specialisation and niche design. The legal industry finds established conventions a useful tool to exercise their trade. The loss comes in preventing systems from going beyond themselves, enabling other systems. The beauty of something like galvanizing is in the possibilities it gives to metalworking itself, thence to a particular sensibility of worked metal. Design and Build procurement is now burgeoning with central government encouragement. A reaction to the consequent blandness and return towards more general consultancies may occur in due course.

More French holidays. A long walk took me down Boulevard Raspail, Paris to see Jean Nouvel's Fondation Cartier. How typical of that architect to juxtapose the jewellery inside with an industrial chic surrounding, a trabeated transparent building of apparent simplicity, planes of galvanized mesh and glass and people hurrying between the wall planes presented in a long elevation to the street. The best place to comprehend the building is from the galvanized fire escape opposite.

Few schools in Britain can boast a footbridge of such architectural distinction as this one. It links the two halves of a girls' school across Plashet Grove, providing traffic-free and weatherproof access. The footbridge meanders for 67 metres in an S-curve at first floor level. The ribbed-tunnel effect is punctured at midspan by two angular steel-and-glass belvederes.

We started working for ourselves and galvanizing appeared here and there in our projects. First of six awards was the 1999 win: The National Glass Centre, Sunderland; incandescent furnaces within, the North Sea sky without, soft light reflected off galvanized grids and gantries; one can almost hear the sound of footfall on metal decking. Simple clip details are used throughout; no hot-working. The zinc-finished components divide into two kinds; those dipped in coating consistencies like on the back of a spoon and those metal-sprayed, bright wire spooling off a reel through a gas flame to splatter onto the metal; another infernal glow.

We back-engineered Jean Prouve's *Maison Tropicale*, exhibited at London's Tate Modern. The great man was notorious for eyeing-in the size of structural components and an assessment was needed before breakfast parties could be held there each day. This was a highpoint in my understanding of galvanizing and sheet metal processing; here was the real McCoy. Few architects understand construction, product design and industrial production as well as Prouve did; tracing through his work was unforgettable.

Jean Prouve's furniture became objects of desire for me. The clothing brand G-Star had reinvigorated the system with denim and called it 'Raw', perfectly complementing the seamed coated metal with an aura of cool. Family holidays commenced and we were sailing gaff rig cutters, (*Swallows and Amazons* boats) at the annual owner's regatta on the Norfolk Broads. The fittings are all crude wrought bits of metal, hot-dip galvanized and let into the teak decks and veneered cabin constructions. We had a roll of Galvanizers Association Awards. The 2000 win was the Workhouse Sports Centre by Proctor and Matthews Architects. The practice carefully detailed industrial materials together, setting metal filigrees, fences and screens amongst the surrounding birches. 2003 saw Sutherland and Hussey's design for Barnhouse awarded, a big back-lot house development in Highgate—a variety of pavilions made up of slender galvanized columns support semi-wrought oak baulks. For the Bellmouth Passage footbridge in Canary Wharf by Patel Taylor Architects, winner in 2006, we made an enclosed crossing using galvanized fascia elements to create a seamless glass and steel enclosure. Next year the housing at Benyon Wharf, Hoxton, designed by JCMT Architects won, a canal-side composition of lofts and apartments with galvanized balconies and access bridges. Carefully detailed to avoid corrosion traps, such lightweight elements are a perfect use of material and protection system. The generic problem of balconies, unmaintained fixings failing under excessive loadings is avoided.

We have judged a few competitions too and the choice of galvanizing and zinc is always set against a background of other materials and systems. The Aluminium Imagination Awards 1999 acknowledged Goldsmiths College use of bright colour anodised metal by Allies and Morrison. It brought to mind Will Alsop's grey galvanized building for the same client nearby: the more playful architect electing to use the more sober material. And then there is zinc in its alloys; brass in cheap earrings, in Victorian firemen's

The prototype house, designed by the French architect Jean Prouve, for 1950s colonial West Africa, is erected outside Tate Modern. Maison Tropicale is an extension of the Design Museum's exhibition. Jean Prouve—*The Poetics of the Technical Object* and the house demonstrates the full scale and vision of Prouve's economy of design.

helmets, Roman shields, that gladiators helmet in Pompeii, the knuckle-dusters on the 'Apache' revolver and in the background of Islamic damascene. It runs like a thread through fine things.

Many of our projects have relied on galvanizing and its use has developed over the years. Plashet Bridge, east London by Birds Portchmouth Russum Architects has exposed industrial grade coatings on its canopy ribs and detail feature. We took the galvanized mullion detail from Bellmouth and enlarged it for a much bigger bridge in Wong Tai Sin, Hong Kong. Quality control of process was an important factor on that project and we learnt a lot more about practicalities. The mainland fabricator of the bridge made welders hand in their torch stubs every day because galvanized fencing wire was being stripped from the fields and used instead of proper rods issued but sold on the black market. Lately we have been using galvanized metal for a new generation of austere and temporary buildings. Junction Arts in Hull takes the city's old bus garage and adds Ward beams and other galvanized components to create a new 'pop-up' venue.

I did some teaching and started writing about engineering. An impressive application came from Bruce Goff using a standard US Navy building system of corrugated galvanized metal sheet and rails. Quonsett huts are based on Iroquis Indian meeting halls. The basic component comprises ribs made up as a pair of angles which, spiked through between, hold the metal cladding. Essential to the idea is the self-healing property of galvanizing. Most recently we have been concentrating on the New Wear Crossing, a bridge just upstream from Sunderland's Glass Centre. The design contrasts secondary steel elements coated in dull matt zinc with primary components in bright glass-flake paint. We are also looking at cathodic protection, something I first saw just five years ago at a dry dock in Newlyn, Cornwall. A fishing boat hull was being scraped down and large zinc ingots bolted on to make sacrificial anodes.

In the bathroom cabinet this morning I noticed zinc pills in a little plastic bottle tricked out to have the patina of zinc carbonate. It is a nice colour, flatter than aluminum oxide. One sees it somewhere in almost everywhere that is manmade.

The Junction Arts and Civic Centre is a 170-seat main auditorium in Goole, East Yorkshire. The building links Goole's two main shopping areas, and creates a covered walkway through to the Victorian market. With its programme of film, theatre, music and dance events, Junction places culture right at the heart of the town and plays a leading role in its regeneration.

What's the matter with matter?

Sarah Wigglesworth, Sarah Wigglesworth Architects

An august and celebrated architect once visited the straw house soon after its completion and commented that it was "crafted". He didn't mean it as a compliment. I remember being a bit insulted at the time, but thinking back on it now, he was probably right. I began to think, however, that what his remark implied was the deliberately unpolished, handmade-ness of our house, the absence of slickness and industrial finish. This is interesting because in actual fact quite a bit of it is made of industrialised products, it is just that these are combined with handmade elements; the appearance is very much not repetitive, mechanical and machine-produced. The handmade elements are the more noticeable because of their increasing rarity in building construction, which, as a rule, tends to employ a limited set of building components and products.

In the ways these materials are employed we were also interested in revealing the making of the building, exposing processes and techniques for all to evidence. This was in deliberate contrast to the commonplace approach to technology which seeks to conceal fixings in its attempt to appear seamless (i.e. 'sophisticated'). Rather, our references were architects such as Otto Wagner and Bill Howell whose bolted stone facades tell the story of their making that engages the mind as much as the eye. We made a deliberate feature of this method in the cloth cladding of our office and in the fixing of the corrugated Ultra sheeting that cloaked the straw bale wall.

Our practice, Sarah Wigglesworth Architects, is proud of its interest in and commitment to the sensory aspects of building construction, stressing the tactile, textural and natural as fundamental signifiers in the language of architecture. This contrasts with the prevalent attention paid to the visual, a situation aided by the various forms of communication media. How a building exists and endures in real space and time is a neglected discourse in architecture, and one that demands to be reclaimed.

People are attracted by difference, and where we have used new materials and interesting juxtapositions of opposites (such as the wattle/steel panels in our front gates) we have had positive reactions. In these cases we were interested in doing something

9/10 Stock Orchard Street (colloquially known as the straw house) is a house and office built on the main railway line from Kings Cross. Built by two architects for their own occupation and exploring living and working on the same site, the project seeks new spatial and material solutions and involves elements of self building. Within a tight budget framework, materials were selected for their low embodied energy, because they performed efficiently for their required function and for their ease of use. The resulting collage of high and low tech includes straw bales, recycled concrete nuggets in wire cages, cloth cladding and cement bags, all features of a new aesthetic approach to ecological architecture.

unconventional that did not come as a pre-packaged product with a performance certificate and a guaranteed lifespan. Indeed, our aim was expressly to avoid this kind of prescription which arises out of concerns for risk and is accompanied by additional costs. If a building is to be as economical as possible this sort of cost penalty must be avoided, and a simplified (crafted) method that perhaps carries other risks is the likely outcome. Our roof meadow is a prime example. Being unable to afford the roof 'system' we were advised was the only way of making it work, we opted for a simple solution involving half the number of layers at about one third of the cost. The roof has survived and thrived through developing its own self-sufficiency; with little maintenance, it has developed its own eco-system in which different species have survived because of their suitability to their particular micro-environment.

In subsequent work (for public as well as private clients) the ethic of longevity is a serious issue and we are usually under obligation to ensure that we are neither profligate with materials nor with energy in use. This means paying much greater attention to fabric energy efficiency, the joining of different products which come with guaranteed performance certification and quality standards; however, the penalty for long life can lead to over-specification. Increasingly we are also required to use details that have been accredited by insurance companies (such as when working with the NHBC). This leaves the designer as someone that effectively samples existing packages of assemblies into a significant collage that constitutes a 'perfect product'. You can appreciate the concerns of the client but this sort of thinking does lead to a greatly narrowed range of products and techniques; one that excludes virtually all new materials (i.e. those without a tried and tested history in use), and especially those eco-materials (which tend to be made of natural materials and, by their nature, weather, need more maintenance and will need replacing at more frequent intervals).

So while there are common threads of interest that bind our work together there are also differences in focus brought about by the unique opportunities offered by each situation. I would even go as far as saying that the greater the constraints, the better we like it. At James Leal Centre (where the previous building had been burnt to the ground by arsonists) and at Cremorne Canoeing Centre, for example, the lack of park invigilation meant the parks in which our buildings sat were subject to vandalism, so we had to find a way of protecting the vulnerable areas of the building (especially windows and doors). In both cases we selected a cladding that was robust and long lasting, using flat galvanised steel sheet (some of which was perforated) at Ray Park and Cor-ten panels at Cremorne. In each building we developed details for window screening in the same material as the main cladding to form a seamless whole. Using sliding shutters so that the building can open and close at Ray Park, and permanent screening to veil the changing room windows at Cremorne, we used a problem to create a beautiful, practical solution.

9/10 Stock Orchard Street: the office wall faces the railway line and to keep the noise of trains to a minimum it is faced in cement bags. The office sits on gabion walls filled with recycled concrete nuggets. This is viewed through the fence which is made of galvanized steel grilles filled with wattle panels.

The James Leal Centre is a visitor centre located in Ray Park, London Borough of Redbridge. Located on the site of the former Ray House which once formed the centrepiece of this landscape, it marks the commencement of a series of green spaces along the Roding Valley Green Grid corridor. The project was intended as a focus of park life, housing a cafe, Park Rangers' offices, training rooms, a music workshop for local youth and an exhibition space. Isolated in the park, the structure is clad in solid and perforated galvanized steel sheets as a defence against vandals.

We believe that necessity breeds invention and we approach every project as an opportunity to research a new material or technology, or develop a new skill or expertise in response to its specific challenges. At Sandal Magna Primary School, an Eco School and one of the lowest predicted carbon schools in the UK, we selected timber as one of the primary structural elements because of its low embodied carbon. But spanning eight metres in timber between the brick cross-walls is beyond the capability of a softwood beam. We asked our QS to carry out a cost comparison between glulam combined with softwood and a solution made entirely of cross-laminated timber (CLT). Although more expensive a product, when factoring the reduced time on-site and the reduction in defects, CLT proved comparable to the alternative, and we were able to argue to our client to use this relatively new product. This way we were able to learn the discipline required of a new product and discover for ourselves what its benefits and drawbacks were. There is no substitute for hands-on experience.

While working within the discipline of a singular technology is comfortable, particularly when learning the rules for the first time, breaking those rules is fun and gives license to invent, especially where money is tight. In our project for the cycle store at Bermondsey Square the structure is a repetitive unit of douglas fir portal frames at 500 millimetre centres (too narrow for a bicycle handlebar to get between), clad with brushed stainless steel triangles secret fixed to them. These were positioned randomly to our instruction to create a pixelated pattern on the outside face of the portals. The magic of this façade is that the direction of the steel brushing means that when the triangles are rotated in their three possible positions they reflect the light differently to create a radiant collage of light.

Procurement methods play an unacknowledged part in determining our material choices. Design and Build, for example, removes the ability to specify the product the architect might like best in favour of generalised descriptions that allow the contractor to tender to their subcontractors (even after commencement on-site) to drive the best bargain. While this might be fine where the product is ubiquitous (plasterboard), a specific effect is hard to achieve where only one company offers the desired product. This limits our vocabulary and dictates what we can say, so is a form of censorship.

Finally, a word about morality. There is a self-imposed and unspoken code of honour in architecture which applauds the simple and minimal while seeing as aberrant the complex and exuberant. Once again, I return to the reaction to our Stock Orchard Street project, which was condemned for having "too many ideas" and having "too many materials". Ignoring the rigorous theoretical underpinnings of our experiment here, we had unwittingly infringed a gospel of architectural morality, which says that editing and reduction is a sign of control. While we would certainly agree that wilful and uncoordinated clashes of gratuitous materiality may be

Wardroper House is a 15-unit affordable housing scheme on St George's Road in Southwark, south London. A glazed brick plinth provides a defence against the noisy street and carries a superstructure clad in cedar boarding, a modern interpretation of the many different styles of housing surrounding the site. The building's structural frame was manufactured from factory-fabricated cold-rolled steel wall and floor panels, which were delivered to the site by lorry and erected by crane. Lightweight and efficient, the entire building structure was erected in a matter of weeks.

Sandal Magna Community Primary School replaced a rundown Victorian primary school in Wakefield, and is one of the most carbon efficient schools in the UK. The school's contemporary design evokes the local architecture of neighbouring red brick terraces and chimneys. Different materials combine to provide texture and interest, and create an extremely tactile building which is also a teaching tool for demonstrating ecological design.

incoherent, we defend the right to resist the contraction of our material language by clients, briefs and procurement systems alike. We prefer both/and to either/or, and if that positions us with the postmodernists rather than the modernists, SWA will continue undeterred in demanding the widest vocabulary possible in expressing our beliefs.

Bermondsey bicycle station provides secure storage space for 76 bikes for local residents and office workers in a new residential and commercial development in Southwark. At the core is a simple structure—timber frames clad on the inside with translucent sheeting to provide shelter, security and natural light. On the outside, stainless steel triangles create a lively and interesting facade that relates to nearby landscaping. Integrated lighting creates jewel-like patterns at night, and provides additional security.

An artist at work
Sophie Ryder

As a child, my mother sent me to jewellery classes on a Saturday morning which is when I first started working with wire. Over the years I kept experimenting, using any kind of wire that I could find including coat hangers. I made scarecrows for the garden, a double decker bus and a 1/4 life-sized jumbo jet! I first started making sculptures at the age of 16 during my Art Foundation course at Kingston Polytechnic. This was followed by a three year Painting diploma course at the Royal Academy Schools and although I loved drawing and painting, it was three-dimensional work that really interested me. Soon after leaving the Academy, I was making large wire sculptures and starting to experiment with bronze. My tutor at the Royal Academy wrote on my diploma that I have "an insatiable urge to create", which sums me up: I am never satisfied, there is always somewhere else to go and something else to explore. I feel compelled to do what I do, like any other kind of artist. Every waking moment of every day I'm thinking about what I'm working on and what I will be making next and I just can't wait to get back into the studio.

I love working on a large-scale, mainly because of the interaction it has with the landscape, both rural and urban. Outdoor sculpture needs to be larger-than-life to have any kind of impact, which prompted me to consider protecting my sculptures from the weather.

It has been approximately 30 years now since I first started using the hot dip galvanizing process. At first it was a novelty for galvanizers who were used to dipping steel girders and gates. They were intrigued to see Minotaurs and hares made from wire. Since then, I think many artists have started using the process. It feels good to have been instrumental in coming up with a new way of making outdoor sculpture. It was exciting for me to be able to work in the medium of my choice and to be able to not only protect the material from the elements but also to make it stronger.

As my work has increased in scale, I now create a sculpture in several sections by working out the exact dimensions according to the size of the galvanizing baths. It is quite an exacting, mathematical process and once the work has been galvanized and delivered to whichever exhibition in whichever country it is going to, the segments are then bolted together on-site. Occasionally I omit

This is one of my largest wire drawings and it is of my daughter Maud's eye. She has very blue eyes and I wanted to try and show that without using any colour. This drawing is made in three pieces because of the weight. Although it looks light, it is very dense in some areas such as the pupil.

the bolting process as I quite like seeing a gap within the sculpture. Whilst I was installing a piece in my garden for the first time, I enjoyed seeing the sky pass through the gap—it gave a feeling of movement to the work, and also made it possible to walk through the piece. When a piece is first galvanized, it comes out of the tank like shiny silver and then after a few weeks the colour tends to dull down which I prefer.

The technique I have developed of making wire sculptures sometimes takes a year to make a large outdoor piece or a 20 foot wire drawing. The length of time seems daunting, but even though I usually know at the start of a piece that it could take months or a year or more, it doesn't seem to put me off! I get such a thrill from experimenting with different materials until I achieve a satisfactory outcome.

In the past few years I have started making maquettes in preparation for the monumental work and then scaling-up several times to get to the final size, otherwise, any small mistake becomes a large mistake when the piece is completed. The small sculptures don't need a very detailed armature as they are less likely to fall over and cause damage, but the larger work needs preparation and thought as to its standing in public places. Instinctively, years of practice have taught me about the balance of a structure and how many points need to be in contact with the ground to prevent toppling.

The armatures for my larger pieces are now very intricate and form part of the overall feel of the piece. The whole shape is there in steel rods before I start adding the wire. It's like a grid onto which I can tie the wire, like a sketch on a canvas before adding the paint. Once I have covered the armature with wire pancakes (as I call them!), the texture and feel of the piece at this stage is soft and transparent. At this point the sculpture goes off to the galvanizer. Once galvanized, the work takes on a completely different persona, it becomes solid and strong and in certain lighting looks like stone or concrete! Yet if back-lit, it changes, and the slightly more fragile transparency returns. In some weather, especially snow, the sculpture can blend into the landscape.

It's wonderful that people enjoy looking at my work and that it touches them emotionally. When I make a piece it is a direct response to an emotion that I have experienced, so it's a truthful expression rather than a manufactured one. I don't set out to impress or shock but sometimes people read into my work and draw their own conclusions. If people go away from one of my exhibitions with a satisfaction that they have a lasting memory of seeing some beautiful work, then that's a bonus for me.

People are always intrigued by why I use animal heads in my work. Well I'm not sure why the Minotaur came about except that I loved Picasso as a child and that I studied Greek mythology at school. It was when making a sculpture of a Minotaur and woman that I decided to come up with a different head for the female body. The hare head immediately came to mind as I visualise the ears as a mane of hair and the bulls head and hare head just

seemed to work so well together. The reason I mask my figures is that I don't want them to be immediately recognisable. I want the onlooker to concoct a story of their own and to imagine the figures to be anyone they want them to be. People wonder if there is a deeper reason why I use the hare head or the bulls head but no, it's as simple as that, they just looked apt. Apart from this, I just don't seem to be able to make human heads! I have tried on many occasions but it just doesn't feel right.

Galvanizing has enabled me to work on a monumental scale and to show my work in sculpture parks and cities all over the world. Without it, the work would be too fragile. My ambition is to make work that can compete with skyscrapers and tall oak trees. To make a piece that looks impossibly strong and monumental, yet fragile and sensitive. Over the next few years I plan to make some even larger pieces and to push the boundaries in terms of movable, transportable sculptures to the limit.

Open Hand was one of the first large-scale sculptures that I made in segments and then left that way rather than joining them together. The challenge and satisfaction of sculpting hands and feet, then scaling up to this monumental size gives me the opportunity to really explore all the little lines, creases and bones.

Acknowledgements

This book would not have possible without the generous help of many individuals, in particular our essay contributors; Sarah Wigglesworth, Matthew Wells of Techniker Ltd, Sophie Ryder and Will Alsop of ALL Design. Special thanks go to Charlie Baker, Juan Dols, Dennis Gilbert and Graham Everitt at VIEW Pictures, Ruth Marler at Arcaid, Alison Catterall at Timothy Soar, Allard Bovenberg, Ian Lawson, Jocelyne van den Bossche, Aneurin Phillips, Ray Wood, Mark Hadden, Craig Auckland, Maud Ryder-Scott and Katherine Holman. We are also grateful of the contributions by ABK Architects, AECOM, Antony Gormley, Austin-Smith:Lord, Barton Engineers, Carmody Groarke, David Chipperfield Architects, Dols Wong Architects, EllisMiller Architects, Had Fab Ltd, Ian Ritchie Architects, JCMT Architects, Kier Construction, LA Architects, Michael Hopkins and Partners, Mills Beaumont Leavey Channon, Nicholas Grimshaw & Partners, O'Donnell + Tuomey, Ramboll, Ray Hole Architects, RIM Fabrications, Sarah Wigglesworth Architects, Snell Associates and Studio Libeskind.

Galvanizers Association is the market development and technical support organisation for the hot dip galvanizing industry. GA has provided free authoritative information and advice to specifiers and potential users since 1949.

Circle Insurance Services Plc is very proud to be associated with Galvanizers Association and the publishing of *The Alchemy of Galvanizing*.

Circle Insurance is an independent Chartered Insurance Brokerage with offices across the UK and with access to Lloyds of London as a placing broker. Circle therefore is able to deliver innovative and bespoke covers that, coupled with high standards of service, sets us apart.

With a very strong presence in the design and construction sector, Circle has a division, Circle Professional Risks, which specialises in Professional Indemnity and Management Liability Insurance for Architects, Designers and Structural Engineers.

Image Credits

Courtesy of Robert Duncan, Nebraska: *Upside Down Kneeling* pp. 4–5

Craig Holmes: pp. 10–11, 106–107, 126

Tim Soar: Catmose Campus pp.13–17

Allard Bovenberg: *Exposure* pp.18–22

Had Fab Ltd: *Exposure* p. 23

Dennis Gilbert (View): Cork Civic Offices pp. 25–29; Garsington Opera Pavilion pp. 38–39, 41–43; Lewis Glucksman Gallery pp. 45–49; Canada Blanch School pp. 117–119

Edmund Sumner (View): Eden p. 31

Iqbal Johal: Eden pp. 32–33, 35–37; Brunel Street Car Park pp. 109–111; 124–125

James Brittain (View): Wardroper House p. 136

Kilian O'Sullivan (View): Eden p. 34

Colin Willoughby: Garsington Opera Pavilion p. 40

Richard Bryant/arcaid.co.uk: Gormley Studio pp. 50–55

Oak Taylor-Smith: Artist's Workshop at Gormley Studio pp. 56–59

Liz Eve: Hengrove Leisure Centre pp. 60–65

Ian Lawson: Imperial War Museum-North pp. 67–71

Ian Ritchie Architects: Leipzig Glass Hall pp. 72–73, 75–76

John Edward Linden/arcaid.co.uk: Leipzig Glass Hall pp. 74, 76–77

Andy Stagg (View): Fawood Children's Centre pp. 79–83

Keith Hunter Photography: Roseisle Distillery pp. 85–89

Mark Hadden Photography: James Leal Centre pp. 91–95, 135; Sandal Magna School p. 136; Bermondsey Bike Shelter p. 137

Aneurin Phillips: Snowdon Visitors' Centre pp. 96–97

Ray Wood Photography: Snowdon Visitors' Centre pp. 98, 100–101

Antenna: Snowdon Visitors' Centre pp. 98–99

Craig Auckland/fotohaus: Benyon Wharf pp. 103, 104–105

Charlie Baker: Homes for Change pp. 113–115

Simon Congdon: Jubilee Campus pp. 120–123

Imperial War Museum: Nissen Hut (H40794) p. 127

Morley Von Sternberg: Lynher Dairy p. 128

Yoshiharu Matsumura: Springtecture H p. 129

Philip Halling: Dutch Barns p. 130

Nick Kane: Plashet School Footbridge p. 131

Dean Nicholas: Maison Tropicale at Tate Modern p. 132

Andy Haslam: Junction Arts and Civic Centre, Goole p. 133

Paul Smoothy: Stock Orchard Street pp. 134–135

Harry Scott: Sophie Ryder's work collection pp. 138–141

Artifice books on architecture
10A Acton Street
London
WC1X 9NG

T +44 (0)207 713 5097
F +44 (0)207 713 8682
sales@artificebooksonline.com
www.artificebooksonline.com

All opinions expressed within this publication are those of the
authors and not necessarily of the publisher.

Designed by Albino Tavares at Artifice books on architecture.

British Library Cataloguing-in-Publication Data.
A CIP record for this book is available from the British Library.

ISBN 978 1 908967 42 8

Artifice books on architecture is an environmentally responsible
company. *The Alchemy of Galvanizing* is printed on sustainably
sourced paper.